VoIP
HACKS™

Other resources from O'Reilly

Related titles
Asterisk: The Future of
   Telephony
Switching to VoIP

Talk is Cheap
Skype Hacks™

Hacks Series Home
hacks.oreilly.com is a community site for developers and power
users of all stripes. Readers learn from each other as they share
their favorite tips and tools for Mac OS X, Linux, Google, Windows XP, and more.

oreilly.com
oreilly.com is more than a complete catalog of O'Reilly books.
You'll also find links to news, events, articles, weblogs, sample
chapters, and code examples.

oreillynet.com is the essential portal for developers interested in
open and emerging technologies, including new platforms, programming languages, and operating systems.

Conferences
O'Reilly brings diverse innovators together to nurture the ideas
that spark revolutionary industries. We specialize in documenting the latest tools and systems, translating the innovator's
knowledge into useful skills for those in the trenches. Visit *conferences.oreilly.com* for our upcoming events.

Safari Bookshelf (*safari.oreilly.com*) is the premier online reference library for programmers and IT professionals. Conduct
searches across more than 1,000 books. Subscribers can zero in
on answers to time-critical questions in a matter of seconds.
Read the books on your Bookshelf from cover to cover or simply flip to the page you need. Try it today with a free trial.

VoIP HACKS™

Ted Wallingford

O'REILLY®

Beijing · Cambridge · Farnham · Köln · Paris · Sebastopol · Taipei · Tokyo

VoIP Hacks™

by Ted Wallingford

Copyright © 2006 O'Reilly Media, Inc. All rights reserved.
Printed in the United States of America.

Published by O'Reilly Media, Inc., 1005 Gravenstein Highway North,
Sebastopol, CA 95472.

O'Reilly books may be purchased for educational, business, or sales promotional use. Online editions are also available for most titles (*safari.oreilly.com*). For more information, contact our corporate/institutional sales department: (800) 998-9938 or *corporate@oreilly.com*.

Editor:	David Brickner	**Production Editor:**	Sanders Kleinfeld
Series Editor:	Rael Dornfest	**Cover Designer:**	Marcia Friedman
Executive Editor:	Dale Dougherty	**Interior Designer:**	David Futato

Printing History:

December 2005:	First Edition.

RepKover™ This book uses RepKover™, a durable and flexible lay-flat binding.

ISBN: 0-596-10133-3
[M]

Contents

Credits ix

Preface xiii

Chapter 1. Broadband VoIP Services 1

    1. Get Connected   2

    2. Use Pure VoIP Dialing with Your TSP   5

    3. Wire Your House Phones for VoIP   7

    4. Use a Softphone with a VoIP TSP   9

    5. Prioritize Packets to Improve Quality   14

    6. Got 911?   17

    7. Update Your VoIP ATA Firmware   19

Chapter 2. Desktop Telephony 22

    8. Access Next-Gen Voice Features   22

    9. Track Vonage Account Info on Your Desktop   24

    10. Pick a Desktop VoIP Client   26

    11. Sound Like Darth Vader While You VoIP   29

    12. Grow Your Social Network with Gizmo   31

    13. Record VoIP Calls on Your Windows PC   35

    14. Handle Calls with Windows Software   36

    15. Let Your Mac Answer and Log Your Calls   40

    16. Run Phlink Even When Logged Off   42

    17. Greet Callers Differently Each Day   43

    18. Use Caller IDs in AppleScripts   44

    19. Control iTunes from Phlink   46

20. VoIP While Fragging 48

21. Google for Telephony Info 50

22. Telephonize a Sound File 52

23. Record an Audio Chat on Your Mac 55

24. Create Telephony Sounds with SoX 57

25. Mix the Perfect Announcement 59

26. Sound Like a Pro Announcer 60

27. Record a Videoconference 61

Chapter 3. Skype and Skyping **63**

28. Get Skype and Make Some New Friends 65

29. Skype Your Outlook Contacts 67

30. Skype People from the OS X Address Book 70

31. Enable Site Visitors to Skype You 72

32. Speak Jyve 73

33. Teach Your Browser to Speak Jyve 76

34. Carry Skype in Your Pocket 78

35. Degunk International SkypeOut Calls 79

36. From Podcasting to Skypecasting 81

37. Answer Your Skype Calls, Even When You're Not Around 84

38. Use Custom Rings and Sounds with Skype 87

39. Emote by Sight and Sound with Skype 89

40. Skype with Your Home Phone 92

Chapter 4. Asterisk **95**

41. Turn Your Linux Box into a PBX 97

42. Attach a SIP Phone to Asterisk 100

43. Connect a Phone Line Using an FXO Gateway 105

44. Connect a Legacy Phone Line Using Zaptel 108

45. Forward Your Home Calls to Your Cell Phone 112

46. Selectively Forward Calls 114

47. Report Telephone Activity with Excel 116

48. Kindly Introduce Telemarketers to Mr. Privacy 122

49. Build a Four-Line Phone Server 124

50. Master Music-on-Hold 129

51. Record Calls 131

52. Get Your Daily Weather Forecast from Your Telephone 133

53. Put a Happy Face on Asterisk Using AMP 133

54. Run Asterisk Without Root, for Security's Sake 137

55. Link Two Asterisk Servers with PSTN 138

56. Link Several PBXs over the Internet 142

57. Route Calls Using Distinctive Ring 145

58. Tune Up Your Asterisk Logs 147

Chapter 5. Telephony Hardware Hacks **150**

59. Record Calls the Old-Fashioned Way 150

60. Make IP-to-IP Phone Calls with a Grandstream BudgeTone 151

61. Build a Custom Ringtone for Your Grandstream Phone 155

62. Tweak Your Sipura ATA 157

63. Build a Bat Phone 162

64. Brew Your Own Zaptel Interface Card 165

65. Build a Speed-Dial Service on Cisco IP Phones 166

66. Power Cisco Phones with Standard Inline Power 169

67. Customize Your Cisco IP Phone's Boot Logo 171

68. Configure Multiple IP Phones at One Time 172

69. Customize Uniden IP Phones from TFTP 177

70. Control the Lights Using Your IP Phone 179

71. Use a Rotary-Dial Phone with VoIP 181

Chapter 6. Navigate the VoIP Network **184**

72. Monitor VoIP Devices 184

73. Inspect the SIP Message Structure 187

74. Audit a Network's QoS Capabilities 189

75. Graph Latency and Jitter 192

76. Explore NAT Traversal 196

77. Shape Network Traffic to Improve Quality of Service 201

78. Create a Premium Class of Service 206

79. Build a $100 PSTN Gateway in 10 Minutes or Less 210

80. Make IP Phone Configuration a Trivial Matter 213

81. Peek Inside of SIP Packets 215

82. Dig into SDP 222

83. Sniff Out Jittery Calls with Ethereal 226

84. Log VoIP Traffic   228
85. Secretly Record VoIP Calls   232
86. Log and Record VoIP Streams   234
87. Intercept and Record a VoIP Call   237

Chapter 7. Hard-Core Voice **241**

88. Build a Killer Telephony Server   242
89. Build an H.323 Gatekeeper Using OpenH323   247
90. Turn Your Linux Box into a Fax Machine   254
91. Build an Inbound Fax-to-Email Gateway   256
92. Teach Your Asterisk Box to Speak   259
93. Build a Mac PBX   262
94. Monitor Asterisk from Your Perl Scripts   266
95. Build a SoftPBX with No Hard Drive   268
96. Build a Standalone Voicemail Server in Less Than a Half-Hour 275
97. Automate Your Voicemail Greeting   278
98. Connect Asterisk to the Skype Network   281
99. Forward Your Home Phone Calls to Skype   283
100. Get Started with sipX   283

Index **287**

Credits

About the Author

Ted Wallingford is a senior network engineer with LCG Technologies Corp. in Elyria, Ohio, and the author of *Switching to VoIP* (O'Reilly). Ted has led installations of VoIP technology in the construction, manufacturing, and networking industries. A periodic contributor to *Macworld* magazine and VoIPfan.com, Ted is a strong advocate of open standards and *Star Wars* movies. He updates the web site *http://www.macvoip.com/* at least a couple of times a year. Ted lives with his wife Kelly and two amazing kids, Jacob and Madelyn, in suburban Cleveland.

Contributors

The following people contributed their writing, code, and inspiration to *VoIP Hacks*:

Brian Degenhardt

Brian's experience in the high-tech industry includes work in such diverse areas as network engineering, online media delivery, and console game development. Currently he serves as CTO of Four Loop Technologies, maker of the Switchvox PBX. Brian has contributed to numerous open source projects, including the GIMP and the Squid web proxy cache. He currently resides in sunny San Diego with his wife Tristan.

Kristian Kielhofner

Kristian is president of KrisCompanies (*http://www.kriscompanies.com/*), a consulting firm based in Lake Geneva, Wisconsin, and creator of AstLinux, a Linux distro configured specifically for Asterisk that features a very small footprint. Kristian has been working with Linux professionally for more than five years, since he began doing Linux system administration at the age of 16. Kristian started KrisCompanies in 2004 to help local businesses with their technology needs. In addition to creating AstLinux, he has also been involved with AstShape and Polycom configuration files.

Andrew Latham

Andrew is a networking consultant who offers VoIP, IP networking, and web development expertise via his web site, *http://www.lathama.com/*.

Dave Mabe

Dave (*http://dave.runningland.com/*) is an accomplished and largely self-taught engineer and writer who strives to create simple, elegant solutions to complex problems. Dave has worked at AT&T in the communications industry for eight years. Always looking to save a few keystrokes and mouse clicks, he is the kind of person who would rather spend several hours inventing an automated solution than spend a few monotonous moments each day performing a menial task. Dave has been using Asterisk and other VoIP solutions for two years. He is the author of *BlackBerry Hacks* (O'Reilly).

Joel Sisko

Joel has been a self-proclaimed network convergence professional since 1992. He is the founder and CEO of Convergence Center LLC, a company focused on delivering the next generation of convergence-based applications and communications systems for value-added voice and data resellers.

Acknowledgments

The contributors who spent their very scarce time working on material for this book have my earnest thanks. These gentlemen are truly talented networking engineers, and their input and technical expertise were invaluable. Do business with these guys—they are established knowledge leaders in this new industry.

I'd also like to thank Mike Loukides, who recommended that I write *VoIP Hacks* (or perhaps he merely succumbed to my nagging). He edited *Switching to VoIP*, and he is the author or co-author of several excellent O'Reilly volumes, including the highly useful *Unix Power Tools* and one of O'Reilly's earliest technical books, *System Performance Tuning*. He's a pretty amazing pianist, too.

VoIP Hacks was edited by David Brickner. David's editing is pragmatic, politically incorrect, and to the point. I love that. I'll give David all the credit for anything good about *VoIP Hacks*!

This book survived the criticisms of several tech reviewers, including Kristian Kielhofner, Leif Madsen, and Jim Van Meggelen. Thank you for looking over my work; your expertise added much to the book's accuracy.

I must heartily acknowledge my hometown crowd, too. My wife Kelly and my friends at Pathway have given me plenty of much-needed encouragement. The crew at LCG Technologies is a great bunch, too. I just barely squeaked this book out thanks to my new workload—keep up the great work, LCG. Thanks to Brian Downey of *The Linux Fix* for his expert Linux support, as well.

Preface

Voice over IP, or *VoIP*, is a family of technologies that enables voice applications and telephony to be carried over an Internet Protocol (IP) network such as the Internet. These technologies include protocols, hardware and software standards, and computer programs. VoIP is employed in telephony applications, from analog phones to next-generation IP phones and wireless headsets, and in desktop voice chat services, from web-based party-line chat services (like Yahoo! Chat) to the well-known Skype desktop voice-calling service.

VoIP has become an important technology that is integrating pervasively into the popular culture. It is employed daily to drive new engines of commerce—everything from business-class VoIP-powered calling services to simple desktop chat tools such as Apple's iChat. Other high-profile companies like eBay, Microsoft, Google, and AT&T offer applications and services that utilize VoIP, too.

These big companies have recognized that the popular culture is moving to VoIP services en masse, even as the telecom industry is being set on its ear by scrappy young VoIP startups like Vonage, Packet8, and SpeakEasy.net. VoIP services deliver telephony applications less expensively than the old phone companies can hope to. This is because VoIP is free of the continually burdensome legacy technology investment the old phone companies must make to keep the "old" global phone network running. VoIP is also free of the endless government regulations and tariffs imposed upon the old phone companies.

In a nutshell, the way society looks at the voice network has changed. VoIP is the enabler of the change, and tomorrow's global voice network is the Internet.

This book contains only a small subset of VoIP knowledge—enough to serve as an introduction to the world of VoIP and teach you how to use it to save money, be more productive, or just impress your friends. My friends love my on-hold music when they call my house; I love that when people call my house, the call is connected to my notebook PC via Skype, no matter where I am in the world. You'll learn how to do all of this and more. I hope this book gets your mental gears turning and that your VoIP hacks are as enjoyable to implement and customize as they were for me to write!

For more VoIP theory and detailed reference information about Voice over IP, check out these great O'Reilly titles:

- *Switching to VoIP*
- *Skype Hacks*
- *Talk Is Cheap*
- *Asterisk: The Future of Telephony*
- *Practical VoIP Using Vocal*

Why VoIP Hacks?

The term *hacker* has a bad reputation in the press. They use it to refer to someone who breaks into systems or wreaks havoc with computers as their weapon. Among people who write code, though, the term *hack* refers to a "quick-and-dirty" solution to a problem, or a clever way to get something done. And the term *hacker* is taken very much as a compliment, referring to someone as being *creative*, having the technical chops to get things done. The Hacks series is an attempt to reclaim the word, document the good ways people are hacking, and pass the hacker ethic of creative participation on to the uninitiated. Seeing how others approach systems and problems is often the quickest way to learn about a new technology.

Since it is based in software, VoIP is overflowing with hack potential. If you love to tinker and optimize, this technology offers a cornucopia of exciting things to tweak and customize. As in the heyday of the World Wide Web, fortunes will be made in the nascent VoIP industry, and lots of fun will be had by voice hackers like you and me.

How This Book Is Organized

You can read this book from cover to cover if you like, but each hack stands on its own, so feel free to browse and jump to the different sections that interest you most. If there's a prerequisite that you need to know about, a cross-reference will guide you to the right hack.

The book is divided into seven chapters, organized by subject:

Chapter 1, *Broadband VoIP Services*

In this chapter, you'll be introduced to some Internet-based VoIP phone service providers who can help you replace your traditional phone line with a cost-saving, feature-rich VoIP line.

Chapter 2, *Desktop Telephony*

Since VoIP is rooted in software, it has some wonderful uses on your desktop PC or Mac. In this chapter, you'll learn how to customize and maximize productivity-enhancing telephony applications.

Chapter 3, *Skype and Skyping*

Skype, the ubiquitous desktop voice-calling application, is one of the most hackable desktop telephony tools, and therefore is worthy of an entire chapter of hacks.

Chapter 4, *Asterisk*

Just as VoIP enables desktop telephony, it also enables enterprise telephony. In this chapter, you'll learn how to install, configure, and hack Asterisk, an open source PBX.

Chapter 5, *Telephony Hardware Hacks*

VoIP is rooted in software, but it is used with lots of different kinds of hardware—everything from next-generation IP phones to old-school rotary phones. This chapter shows you how to add these devices to your VoIP setup—and how to customize them.

Chapter 6, *Navigate the VoIP Network*

Voice over IP is carried over the network using packets, just like traditional data. With the advice in this chapter, you can monitor VoIP and troubleshoot it using traditional admin tools.

Chapter 7, *Hard-Core Voice*

By the time you reach this chapter, you will have advanced to the hallowed ground that's held by a very exclusive crowd: the community of hard-core voice hackers.

Conventions Used in This Book

The following is a list of the typographical conventions used in this book:

Italics

Used to indicate URLs, filenames, filename extensions, and directory/folder names. For example, a path in the filesystem will appear as */Developer/Applications*.

Constant width

Used to show code examples, the contents of files, console output, as well as the names of variables, commands, and other code excerpts.

Constant width bold

Used to show user input in code and to highlight portions of code, typically new additions to old code.

Constant width italic

Used in code examples and tables to show sample text to be replaced with your own values.

Gray type

Used to indicate a cross-reference within the text.

You should pay special attention to notes set apart from the text with the following icons:

This is a tip, suggestion, or general note. It contains useful supplementary information about the topic at hand.

This is a warning or note of caution, often indicating that your money or your privacy might be at risk.

The thermometer icons, found next to each hack, indicate the relative complexity of the hack:

 beginner   moderate   expert

Using Code Examples

This book is here to help you get your job done. In general, you may use the code in this book in your programs and documentation. You do not need to contact us for permission unless you're reproducing a significant portion of the code. For example, writing a program that uses several chunks of code from this book does not require permission. Selling or distributing a CD-ROM of examples from O'Reilly books *does* require permission. Answering a question by citing this book and quoting example code does not require permission. Incorporating a significant amount of example code from this book into your product's documentation *does* require permission.

We appreciate, but do not require, attribution. An attribution usually includes the title, author, publisher, and ISBN. For example: "*VoIP Hacks* by Ted Wallingford. Copyright 2006, O'Reilly Media, Inc., 0-596-10133-3."

If you feel your use of code examples falls outside fair use or the permission given above, feel free to contact us at *permissions@oreilly.com*.

Safari® Enabled

 When you see a Safari® Enabled icon on the cover of your favorite technology book, that means the book is available online through the O'Reilly Network Safari Bookshelf.

Safari offers a solution that's better than e-books. It's a virtual library that lets you easily search thousands of top tech books, cut and paste code samples, download chapters, and find quick answers when you need the most accurate, current information. Try it for free at *http://safari.oreilly.com*.

How to Contact Us

We have tested and verified the information in this book to the best of our ability, but you may find that features have changed (or even that we have made mistakes!). As a reader of this book, you can help us to improve future editions by sending us your feedback. Please let us know about any errors, inaccuracies, bugs, misleading or confusing statements, and typos that you find anywhere in this book.

Please also let us know what we can do to make this book more useful to you. We take your comments seriously, and we will try to incorporate reasonable suggestions into future editions. You can write to us at:

O'Reilly Media, Inc.
1005 Gravenstein Highway North
Sebastopol, CA 95472
(800) 998-9938 (in the U.S. or Canada)
(707) 829-0515 (international/local)
(707) 829-0104 (fax)

To ask technical questions or to comment on the book, send email to:

bookquestions@oreilly.com

The web site for *VoIP Hacks* lists examples, errata, and plans for future editions. You can find this page at:

http://www.oreilly.com/catalog/voiphks

For more information about this book and others, see the O'Reilly web site:

http://www.oreilly.com

To reach the author of this book, Ted Wallingford, you can send an email to:

ted@macvoip.com

Got a Hack?

To explore Hacks books online or to contribute a hack for future titles, visit:

http://hacks.oreilly.com

Broadband VoIP Services

Hacks 1–7

Voice over IP (or VoIP for short) is a technology that allows Internet Protocol (IP) networks like the Internet to be used to enable voice communication, similar in some ways to a telephone. Some folks call VoIP *IP telephony*—and the technology comes in many forms, from desktop communication software to automated message recording and fax integration tools.

But in its simplest form, IP telephony enables you to place phone calls over the Internet rather than over a traditional phone line. This is a pretty big deal, since no long-distance charges or hefty federal access taxes are levied on Internet-based phone calls. Plus, IP telephony lets you integrate your desktop PC, your desk phone, and your cell phone in ways never before imagined. I'm anxious to share the details with you in this book.

In the tradition of O'Reilly's Hacks book series, you'll be using short hacks, like the basic ones in this chapter, to learn about Voice over IP and computer-based telephony. I, and a number of my peers in the telecommunications industry, have contributed some of the most useful, most educational, and coolest projects to *VoIP Hacks*. Hopefully, beginning right here in this chapter, you'll be saying, "I didn't know you could do *that* with VoIP!"

VoIP-Based Phone Service Providers

The Golden Age of broadband began with catchphrases like "surf the Web five times faster" and with promises of ultra-fast music downloads. But in the late 1990s, few would have predicted that VoIP-based telephony would be one of the biggest beneficiaries of once-hyped broadband technologies like cable Internet and DSL. Sure, web surfing at "the speed of light" and downloading music are great—but can they save you money? Legally?

VoIP telephony can—and does. For roughly half the cost of a traditional phone line, you can subscribe to a VoIP telephony service provider rather than to a phone company. You'll get a standard phone number that people from the non-VoIP world can use to call you—and you won't have to pay $5 a month extra for voicemail and caller ID.

This chapter has a handful of hacks that will show you how to maximize your broadband voice service. So, if you subscribe to a VoIP service provider, you're ready to hack. If not, what are you waiting for? "Get Connected" [Hack #1] describes some VoIP-based phone providers that you should evaluate as you prepare to dive into *VoIP Hacks*.

Get Connected

If you've got broadband, you're already using the Internet for data communication. Wouldn't it be great to use it for telephone calls, too?

Internet telephony service providers (TSPs) get your voice onto the Net, allow you to make and receive phone calls just like traditional phone companies, and tend to shrink your phone bill to boot. Some of these service providers give you a basic, free service that enables you to call other users over the Internet. Others allow you to make toll-free calls free of charge, but charge for local and long-distance calls.

TSPs that allow you to call traditional telephone service subscribers do so by connecting your standard home phone to the Net. Some TSPs also let you use a special piece of software called a *softphone* to place calls with your PC. To get connected to a TSP, you need a broadband Internet router configured as a DHCP server, a spare Ethernet port either on your router or on a nearby switch, and a good old-fashioned analog telephone.

TSPs are data centers with telephony servers that route calls to and from your home network or broadband VoIP device. The real-time packets that carry each call's sound over your broadband link use IP and User Datagram Protocol (UDP) protocols, and the TSP communicates key moments in the call—like dialing, connecting, and hanging up—using signaling protocols that are similar in some ways to the ones your browser uses to surf the Web.

The VoIP device that most TSPs provide to connect your home phone is known as an *analog telephone adapter*, or ATA. These little boxes allow you to connect a residential-style analog phone to your broadband Internet connection, and they are normally supplied by your VoIP TSP when you sign up for their service.

In addition to an ATA, some TSPs permit you to place VoIP calls using the following:

An IP phone

These telephones connect directly to an Ethernet network using a patch cable or wireless link. They have an IP address as a PC would, and they communicate with the VoIP TSP's data center over your Internet broadband link.

A softphone

These are software programs that run on a PC and permit telephone-style communication using your broadband link. They appear on your Windows, Linux, or Mac desktop with graphical user interfaces that often resemble a telephone, and they require that your PC have a microphone and speakers.

For this hack, I'll concentrate on connecting to a TSP that provides an ATA, allowing you to use an analog phone to place and receive calls via the Internet. Table 1-1 lists domestic (U.S. and Canada) TSPs that provide broadband VoIP calling.

"Bring your own device" means the TSP allows you to make phone calls across its VoIP network using your choice of equipment, such as an IP phone, a PC, or your own ATA. TSPs that don't allow you to bring your own device will provide an ATA to make the connection.

Table 1-1. VoIP TSPs

Company	Web site	Bring your own device
AT&T CallVantage	*http://www.att.com/*	No
BroadVoice	*http://www.broadvoice.com/*	Yes
Broadvox Direct	*http://www.broadvoxdirect.com/*	No
Net2Phone	*http://www.net2phone.com/*	No
nikotel	*http://www.nikotel.com/*	No
Packet8	*http://www.packet8.net/*	No
SOYO	*http://phone.soyo.com/*	No
VoicePulse	*http://www.voicepulse.com/*	Yes
Vonage	*http://www.vonage.com/*	No

Once you've subscribed to a VoIP TSP service (many allow you to subscribe on the Web) and you've received your ATA in the mail, you'll probably be itching to hook it up and use it. Most of the time, setting up an ATA is straightforward. All ATAs have an Ethernet interface, for connecting to your network via an eight-wire CAT5 patch cable with two RJ45 connectors, and one analog telephone interface, for connecting to a residential-style, single-line phone using a four-wire patch cable with two RJ11 connectors. 8x8 Inc.'s DTA-310, standard equipment for Packet8 service, is such a device. So is the Sipura SPA-2000 [Hack #62], pictured in Figure 1-1.

Figure 1-1. The front and rear panels of the Sipura SPA-2000 ATA

But other ATAs might offer additional capabilities. For instance, the Sipura SPA-2000, standard equipment for VoicePulse service, offers an extra analog phone connector, so you can easily connect two phones, or perhaps a phone and an answering machine. As shown in Figure 1-1, the SPA-200's front panel has two phone connectors and a status LED, which indicates whether one of the analog phones is off the hook. The rear panel has an Ethernet connector, an Ethernet activity/link indicator LED, and a DC power connector.

More elaborate ATA devices integrate broadband routing and firewall functions, allowing you to consolidate your VoIP ATA and residential firewall into a single unit. The Zoom 5567 is one of these. It has a broadband IP router with a firewall, a four-port switch, an analog phone connector, and a pass-through connector for placing calls on a traditional Bell phone line in the event the Internet service fails.

So, depending on your service, setup could be a little more elaborate than just connecting the phone and the Ethernet to your ATA. However, in most cases, the ATA is a simple, no-frills device designed to accomplish one thing—get your analog phone connected to the world's biggest VoIP carrier network, the trusty ol' Internet.

After you've gotten the ATA out of the package, find a good place for it. It should be close to where you intend to use the analog phone, though a long-enough phone cord would afford you more distance. (In "Wire Your House Phones for VoIP" [Hack #3], you'll see how to use your house's existing phone wiring to hook up several phones to a single ATA.) Your ATA also needs to be close enough to your Ethernet switch or broadband router to connect to it with a CAT5 patch cable.

Once connected, most ATAs will automatically register with your VoIP service provider's server the first time they are powered up. Don't interrupt this process. If the initial registration is interrupted, it could render your ATA useless, and the TSP might need to exchange it for a new one. Some ATAs will download firmware patches during the initial registration, too. Refer to your ATA's instructions for indications on when this process is complete, making it safe to power off the ATA. Usually, if you can hear a dial tone on the connected phone, the process is complete and it's safe to place a call or power down the ATA.

If you can't hear a dial tone on the connected phone, check that it is connected to the appropriate port on the ATA. Make sure your broadband router is configured as a DHCP server. Without DHCP running on your network, the ATA will be unable to obtain an IP address, crippling it.

> Many VoIP calling plans require that you dial the full 11-digit phone number, even if you're just calling your next-door neighbor. So, if you make a lot of local calls, get used to dialing your own area code a lot!

Once you hear a dial tone, it's probably best to investigate any features that are included with your calling plan—voicemail especially. Then, try calling a buddy to see if you can hear any difference between a traditional call and a VoIP one. Chances are that the person on the other end won't notice the difference unless you tell him you're on a VoIP call. Then, he might say he suspected you were on a cell phone. The sound quality on a VoIP call is only as good as the network carrying it, and many unsuspecting participants mistake VoIP calls for cell phone calls.

HACK #2 Use Pure VoIP Dialing with Your TSP

By using dialing shortcuts, you can keep your phone calls on the Internet and avoid extra charges.

If you're able to make a phone call to a regular phone company subscriber using your new VoIP service [Hack #1], you're ready to learn some cool TSP tricks.

Your VoIP phone bill is probably lower than that of your friends who still use traditional calling plans. But a lower phone bill isn't the only luxury that comes with converting your service to VoIP. Because your call uses the Internet rather than the public telephone network to route your call, you have access to several cool dialing shortcuts when you call subscribers of other

VoIP services. When an IP network alone provides the pathway between caller and receiver, it's said to be *pure* (or *native*) Voice over IP.

This can actually save you money, especially if you make a lot of international calls. If you're a Free World Dialup (FWD) subscriber and you talk frequently with your buddy in Mexico, who uses Vonage, using dialing shortcuts will keep your calls pure VoIP and allow you to circumvent any related long-distance calling charges that would be assessed if your calls were to traverse the Public Switched Telephone Network (PSTN).

To make pure VoIP calls using your TSP's service, you have to be aware of the dialing shortcuts your TSP provides to route calls to other TSP networks using the Internet—instead of the PSTN—as the carrier network. Most VoIP TSPs will assume your call is destined for the PSTN—just because it's an 11-digit phone number. So these shortcuts tell the TSP that you don't want to route your call to the PSTN. Instead, you want to route it over the Internet to another VoIP TSP.

Why do this? If you have an unlimited calling plan, it won't really save you any money. The call probably won't sound any better either. But this technique does conserve your TSP's public telephone network capacity when you use pure VoIP rather than VoIP-to-PSTN calling. If your VoIP TSP bills you by the minute, it might not charge for calls that don't use its PSTN capacity. Plus, it's just cool to let the Internet replace the Bell System for your phone calls. Here's how.

VoIP services such as FWD, Vonage, IAXTel, VoicePulse, and Packet8 offer dialing shortcuts to allow calls between their customers. If you're a Packet8 subscriber, you can reach any FWD subscriber by dialing 0451 and the six-digit FWD number assigned to that subscriber (FWD subscribers don't have traditional 11-digit phone numbers because the service doesn't provide PSTN calling). Consult Table 1-2 for a rundown of the VoIP dialing shortcuts that you can use to route calls between the various VoIP services.

Table 1-2. Pure VoIP dialing between TSPs using the Internet

Desired action	Dialing shortcut
Call an IAXTel user from FWD	*-1-700 and the seven-digit IAXTel number
Call a Vonage user from FWD	**-2431 and the full 11-digit Vonage PSTN number
Call an FWD user from Vonage	0110393 and the six-digit or five-digit FWD number
Call an FWD user from Packet8	0451 and the six-digit FWD number or five-digit FWD number
Call a Packet8 user from FWD	**898-1 and the full 11-digit Packet8 PSTN number
Call a VoicePulse user from FWD	1-700-900-0000 and the full 11-digit VoicePulse PSTN number

Table 1-2. Pure VoIP dialing between TSPs using the Internet (continued)

Desired action	Dialing shortcut
Call an IAXTel user from VoicePulse	1-700 and the seven-digit IAXTel number
Call an FWD user from VoicePulse	1-700-9 and the six-digit FWD number, or 1-700-99 and the five-digit FWD number

Wire Your House Phones for VoIP

You can use your home phone wiring to connect all your home phones to your VoIP service.

If you're happy with your VoIP service, you might want to consider replacing your existing land-line telephone service with that of your new VoIP TSP. This means you must provide a dial tone to all of your analog phones using the ATA instead of a connection from the phone company. Your problem is that most ATAs have only a single analog phone connector, limiting you to just one phone. Radio Shack sells two-wire phone splitters that you can use to connect two analog phones to the same jack—such as the one on your ATA—but this isn't an ideal solution. Who wants telephone patch cables snaking across the floor, anyway?

> Emergency 911 service is required on all VoIP lines sold in the U.S. But since VoIP TSPs handle emergency call routing differently than the old Bell system, it's best to check with your TSP to determine how they handle 911 calls. This way, you'll know what to expect should you need to dial 911.

Fortunately, you already have all the wiring you need throughout your house to share a single VoIP provider's service with multiple analog phones. The phone wiring in most homes is a two-wire or four-wire cable that runs from the telephone company's point of entry, called the *demarc*, to various rooms in the house. In these rooms, a standard modular phone jack provides a place to connect a phone using an RJ11-equipped telephone patch cable. Modular jacks can support up to two phone lines, since analog residential telephony requires only two wires per line. The vast majority of telephone company subscribers use only a single phone line, though.

The analog wiring in the home provides a single-loop parallel circuit, which means that you can piggyback modular jacks off each other. If you need to connect a phone in a new room, you just locate the nearest modular jack and run the wiring to it, instead of running the wiring from the new room all the way to the demarc. In the same way, you can connect the ATA to any modular jack in the house, and all of the analog phones connected to the other jacks will be able to use the service provided through the ATA.

Before you do this, however, it's very important to disconnect the wires from the phone company at the demarc, because the electric current supplied over the phone-company lines could damage the ATA. It's best to find the demarc while your phone company service is active—that way, you can hear the dial tone disappear when you've disconnected the right pair of wires at the demarc. Find your demarc, usually a gray or brown box mounted on the exterior of the building. Inside the box is a cross-connect terminal with screw taps. On one side of the terminal are the wires going into the building. On the other side are the wires from the phone company.

Carefully disconnect the wires from the phone company; the dial tone on the modular jacks inside the building should disappear. (You can take a cordless phone with you to the demarc to listen while you're working.) Even if your phone company lines are dead—that is, you have no phone company service—it's still a good idea to disconnect them. Disconnecting the wires from the phone company side of the demarc will prevent electrical damage to your ATA in the event the phone company turns the lines back on by mistake.

> Don't accidentally disconnect your DSL line! If you have DSL Internet access from the phone company, you might not be able to disconnect your phone company voice service without inadvertently severing the DSL connection, too. Sometimes DSL runs on the same pair of wires as a traditional analog phone service. If you have DSL, it must be on a separate pair of wires from your voice phone line, or this hack won't work, and you will have disconnected your Internet service to boot! Cable Internet subscribers can hack without this worry.

If you're attempting this hack in an apartment, it might be a little tougher. Your lease agreement might prohibit you from making wiring changes like this. In some jurisdictions, the phone company itself or building codes might prohibit this type of wiring hack. If you're not sure, call the phone company and ask that a lineman come out to disconnect the wiring.

Once disconnected from the demarc, mark the pair of wires with a tag that reads, "Phone company: Do not reconnect." This will prevent a well-meaning phone company service technician from reconnecting your line and frying your ATA.

Assuming your phone company disconnect was successful, you can now connect the ATA into any RJ11 modular jack on the premises. This will let you hear the dial tone generated by the ATA and make VoIP calls through any phones that are connected to the other modular jacks throughout your home.

Most ATAs are designed to handle the power requirements of only a phone or two, so check with the manufacturer of your ATA to see if you can reliably connect more. I have two analog phones and two cordless phones (which receive their power separately, anyway) connected to an 8x8 DTA-310, and I don't experience any problems.

> Devices that use analog modems to communicate on traditional phone lines, like older TiVo boxes and fax machines, can't be used with the analog service provided by an ATA. The fault lies with the analog-to-digital conversion of VoIP codecs, not with the modem itself.

Use a Softphone with a VoIP TSP
Get started with prevalent and freely available SIP softphones.

Depending upon which TSP you choose for your broadband VoIP service, your service agreement might limit you to using only analog phones connected to an ATA. However, if you have a lenient Bring Your Own Device (BYOD) service agreement, your TSP will allow you to use your choice of IP telephony access devices. This might mean you can use an IP phone, a PC softphone, an ATA of your choosing, or even your own telephony server (Chapter 4 is dedicated to this proposition) with the TSP's service. This hack will show you how to use Counterpath's X-Lite softphone with your TSP. But first, a little background on telephone networks, both analog and VoIP.

When you subscribe to broadband VoIP service, what you're really doing is buying a single pathway through the TSP's network. Likewise, when you subscribe to traditional phone service, you're really just leasing a telephone line. With that line, you can use cordless phones, corded analog phones, answering machines, fax machines, modems, and all kinds of other access devices. These different analog devices all use the same electrical access signaling to communicate with the phone company. You could think of this analog protocol as even more primitive than the Morse code. It's simple, but it's what allows analog phone devices to place and receive calls.

If legacy telephony devices are more primitive than the Morse code, Session Initiation Protocol (SIP), the predominant VoIP access signaling protocol, is light-years ahead of both. SIP is a suite of media-signaling software specs that define how streaming media devices (and applications) should interact.

The most significant of modern streaming media apps is IP telephony, of course, which brings me to my point.

Unlike old-fashioned telephone signaling, which is Plug and Play (PnP), using a softphone is a bit more involved. To understand how a softphone works (or an ATA or IP phone, for that matter), you must have a simple grasp of SIP. Although SIP is a sprawling specification with dozens of proposed spinoffs and major revisions, you need to know only a few things to get by with a SIP softphone.

SIP is a lot like Simple Mail Transfer Protocol (SMTP). If you're comfortable with that, SIP will make a lot of sense to you. Like SMTP, SIP clients (the phones) send packet messages to SIP servers (such as proxies and telephone systems) or to other SIP clients (such as other SIP phones). In these packet messages are *headers*, strings of data that form requests for specific functionality from the device on the receiving end. The requests could be to establish a phone call, or merely to let a SIP server know that the phone making the request is available to receive calls. Another function of these requests is authentication. On many systems—like your broadband TSP's VoIP network—the calling device must register and pass a username/password authentication to place or receive calls.

Different TSPs, Different Policies

SIP softphones, such as CounterPath's X-Lite, have many, many built-in features. They can signal call transfers, place callers on hold, and even do conference calling so that three or more parties can talk together. But whether these features are enabled by your TSP is another issue. To conference-call, for example, you might need to pay for an extra "line." Bear in mind that from one VoIP service provider to the next, even a feature-heavy softphone product could be impotent (and then there are those TSPs, such as Packet8, that don't support softphones at all).

Install the Softphone

To get X-Lite, download it from *http://www.counterpath.com/*. X-Lite is, in fact, a scaled-down freeware version of X-PRO, but for the purposes of this hack, the feature disparity between versions makes no difference. Installation is straightforward. On Windows, run the installer package, and on the Mac, drag the X-Lite program icon into your Applications folder. Once installed, launch X-Lite, step through its Audio Tuning Wizard, and look at its user interface. By some strange coincidence, it resembles a nice-looking business phone. Imagine that.

Vonage Users, Beware

If you're a Vonage subscriber, you can download the Vonage-branded version of X-Lite's commercial counterpart, called X-PRO, from your account page on Vonage's web site. If you're using Vonage, you're limited to using Vonage's version of X-PRO, and you won't have nearly the flexibility that the non-Vonage version of the software provides. Indeed, once you have the Vonage version running, the only administrative customization you can do is to change your username and password. You can't really get at the softphone's SIP guts because Vonage's version keeps all that stuff off-limits to the end user. Those seeking a deep hacking experience should probably consider BroadVoice instead. Unlike Vonage, BroadVoice openly supports noncrippled softphones such as X-Lite.

Setting up the basics. After you've gotten through X-Lite's Audio Tuning Wizard, you're ready to dive into the SIP configuration settings. These define how the softphone will authenticate and interact with your TSP's SIP server. To access X-Lite's configuration settings, click the button to the right of the CLEAR button on X-Lite's main window, as shown in Figure 1-2.

Figure 1-2. X-Lite's main window looks a bit like a cellular phone

When the configuration window appears, double-click System Settings, and then double-click Network This to bring up the network configurations (Figure 1-3). Find the Provider DNS Address setting and change its value to the DNS server provided by your VoIP TSP (not your Internet Service Provider, or ISP). Your VoIP TSP might require the use of its own DNS because its SIP resources might be on a private domain that cannot be resolved through the public DNS system. If your VoIP TSP didn't provide a DNS address, you can leave this setting blank.

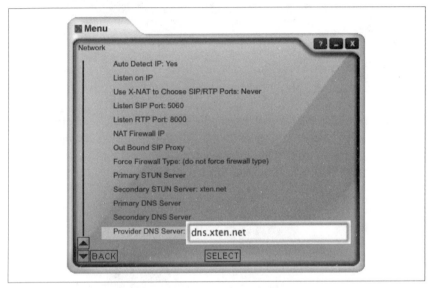

Figure 1-3. X-Lite's network configuration window

Click the Back button in the lower-left corner of the screen to get back to the prior window. Here, you'll need to double-click SIP Proxy to open the SIP Proxy Settings window. Double-click Default, and you'll be able to configure the softphone to use a SIP proxy server, which is located at your VoIP TSP and routes your softphone's calls. The X-Lite softphone can use more than one SIP proxy, but in most situations, you'll need to use only one. This list describes the settings you will need to configure:

Display Name
    Your name, or as much of it as you can fit.

User Name
    The SIP username provided by your TSP. This is likely to be your phone number, including the area code.

Authorization User

> This is normally the same as the SIP username provided by your TSP, though some TSPs might issue a distinct authorization username. In circumstances where multiple phones with their own phone numbers are authorized for the same subscriber, the two usernames might vary.

Password

> The password you and your TSP established when you set up your VoIP account.

Domain/Realm

> This tends to be the domain name associated with your SIP user URI, which is similar to an email address. For *4403281414@sip.broadvoice. com*, your domain/realm would be *sip.broadvoice.com*. Your TSP will issue you an appropriate realm name if it supports the use of softphones.

SIP Proxy

> This is the address of the proxy server that will handle all your VoIP registration activity—things like authentication and notifying the server that you're available to receive calls. Unlike a SIP URI, which always contains *sip* in the domain, the SIP proxy address can be any valid hostname. Again, your TSP will provide you the address to use when you sign up.

Outbound Proxy

> This address is used to handle SIP requests that are bound for other SIP domains. Since most VoIP TSPs don't support calling other realms using SIP, and generally only support calls to and from the public telephone network, an outbound proxy isn't necessary. But BroadVoice, for one, requires that you configure an outbound proxy address—and in its case, it's the same address as that used for the SIP proxy setting.

Use Outbound Proxy

> This setting tells X-Lite whether you want to treat all SIP requests as though they are destined for another realm. This effectively circumvents the SIP proxy for any activity other than registration, though if the two proxies have the same address, as in BroadVoice's configuration, it doesn't matter what this setting is set to. The choices are Always and Never, in case you were wondering.

> Some TSPs have more than one SIP proxy, and they might allow you to choose among them. To determine which one to use, ping them all. The one with the least amount of variance from one ping packet to the next is the one you want.

Register

> This setting tells X-Lite whether you want the SIP client to authenticate and register with the SIP proxy server. It's very uncommon *not* to register, and you won't get very far with your VoIP service if you don't. So definitely set this one to Always.

For most TSPs, you can leave the rest of the settings unchanged. For a more detailed description of X-Lite's settings, you can download a PDF user manual from CounterPath's web site, *http://www.counterpath.com/*.

Make the Call

When the X-Lite phone has successfully registered with the TSP's proxy, its main window will display a message like "Logged In—Enter a Phone Number." Now, you should be able to type in a valid public telephone network number (try your cell phone for an easy test, if you have one). The service should function at least as well as it would via an ATA and analog phone, with one possible exception—echo. Echo is common with softphones if you're using your PC speakers to listen to the person on the other end of the call. If you experience echo when you speak, use a pair of headphones to cancel the acoustic feedback loop.

HACK   # Prioritize Packets to Improve Quality

#5   Voice traffic competes for available bandwidth on your broadband connection. If there is not enough bandwidth, packets get dropped.

VoIP media streams require a constant, uninterrupted data flow. This data flow is composed of UDP packets that each carry between 10 and 30 milliseconds of sound information. Ideally, each packet in a media stream is evenly spaced and of the same size. In a perfect world, a packet never arrives out of sequence or gets dropped. Voice over IP media packets are framed in a highly precise, performance-sensitive way, described in more detail in *Switching to VoIP* (O'Reilly). Dropped packets and *packet jitter* (packets arriving out of order) cause problems—big problems—for an ongoing call. These problems can cause the voices on the call to sound robotic, to cut in and out, or to go silent altogether.

Most of the packet-drop problems you'll encounter while VoIPing will be the fault of your bandwidth-limited ISP connection—the link from the ISP's network to your broadband router. If you're downloading songs to your iPod, surfing the O'Reilly Network, and patching your *World of Warcraft* client all at once, you won't have enough bandwidth left over to support a VoIP call, but there's a way to curb all those applications' thirst for bandwidth so that you can still VoIP successfully. Read on.

To maximize call quality, the network connection carrying VoIP media packets must be as reliable and consistent as possible. The data link to the ISP should treat all voice media traffic with *high priority*. That is, a VoIP packet gets handled first, as it is more important than another packet—say, for your BitTorrent upload. If the data link is swamped and is out of capacity to carry any more data, less important packets are discarded before more important ones. The net result—for high-priority services like voice—is better Quality of Service, or QoS. Several standards exist to ensure that QoS can occur in a broadband VoIP setup, chief among them: Type of Service (ToS) and 802.1p.

If your broadband router is relatively new, it might support these standards—so enabling packet prioritization is just a matter of flipping some configuration switches.

Prioritize Packets on a Linksys Broadband Router

ToS is a feature of Ethernet switches that permits packets tagged as high priority to be handled first, maximizing their QoS. 802.1p is a similar concept, but tends to hang around on routers, not switches. The Linksys BEFSR81 broadband router is a device that supports 802.1p. It sells for less than $100 USD online, and you can probably find one secondhand on eBay for even less.

In fact, setting up priorities on this router is a snap, thanks to Linksys's usual snazzy web-based interface. Once you get the router unboxed and hooked up, use the web interface to locate the QoS screen. (You'll see it after you click on the Advanced Configuration button and the QoS tab.)

The QoS screen contains two sections: one that allows you to establish queuing priorities for packets depending on their TCP/UDP port numbers, and one that allows you to alter the queuing priority depending upon which Ethernet switch port the traffic originated from. That is, since this router has a built-in switch, you can prioritize some of its eight Ethernet ports using the lower half of the QoS screen.

Prioritize RTP traffic. Most VoIP media streams are carried by Real-time Transport Protocol (RTP) packets. To raise the priority of RTP traffic, enter the port numbers 5004 and 5005, each on its own line, in the section labeled "Application-based QoS," and click on the High Priority radio button for each. After restarting the router, all RTP traffic sent from the router will be handled before any other traffic. This technique is especially good if your LAN has multiple VoIP devices that send media streams through the router.

Prioritize all the traffic from your VoIP ATA. If you have only a single VoIP device to support, like a TSP-provided ATA, it might be best if you tell the router to prioritize traffic by Ethernet port instead of by application, as in the preceding paragraph. Specifically, you want your router to prioritize traffic that comes from the Ethernet port where your ATA is connected. To do so, use the High Priority and Low Priority radio buttons for the numbered Ethernet ports. Set them up however you want and reset the router.

Prioritize all the traffic from an attached Ethernet switch. By setting the priority of a particular Ethernet port, you are telling the router to prioritize anything from the device connected on this port, even if this device is another switch. So, an easy way to give priority to all your dedicated VoIP devices, like IP phones and ATAs, is to connect them all to the same switch and then connect that switch to a high-priority Ethernet port on the router.

Prioritize Traffic on a Standalone Switch

Many workgroup Ethernet switches offer QoS features that used to be found only on advanced "managed" switches. These days, inexpensive switches like the NETGEAR GS605 provide support for ToS and 802.1p. By placing such a switch between your broadband router and your VoIP device, with voice traffic prioritized, you can ensure that outbound voice streams get sent to your broadband router before anything else.

What Happens When VoIP Passes Your Router

Unfortunately, no matter how well prioritized and orderly your VoIP media traffic is when it's forwarded by your broadband router, it still might get slowed down, ripped up, and otherwise tattered as it makes its way across the Internet. The same is true of media packets that come from the Internet to your router—the packets carrying the voice of the person speaking to you. Since you're *receiving*—not transmitting—those packets, you can't really prioritize them. That's the responsibility of the routers that carried the packet to your router—and many routers on the Net these days are ignorant of QoS.

In short, you can control traffic sent from your network, but not traffic sent from other networks to yours. At first blush, this sounds like a threat to broadband VoIP, but over the last few years, many have discovered that the outbound traffic is all you really need to prioritize to have success with a broadband TSP. This is because most broadband ISPs limit the amount of outbound bandwidth available to each customer to discourage customers from hosting high-traffic services on their residential broadband connections.

So, there's less available bandwidth to you for sending than for receiving. The VoIP media stream most likely to suffer as a result is the outbound stream, the one carrying your voice to the person on the other end of the call. As such, it's appropriate to prioritize outbound traffic to overcome the limits many ISPs force on outbound bandwidth.

Got 911?

For a multitude of technical and political reasons, Internet TSPs have been slow to make reliable Emergency 911 dispatch dialing available for their customers. Here's how to know if you've got it.

If you recently signed up for VoIP telephone service, the likelihood of you having 911 service is low, but some TSPs do offer it. The fastest way to find out if your TSP offers it is to contact them and ask. Vonage, for instance, supports 911 call routing to most public safety jurisdictions, but you've got to activate this "feature" first. Here's a snippet from Vonage's end-user agreement:

> You acknowledge and understand that 911 dialing does not function unless you have successfully activated the 911dialing (sic) feature by following the instructions from the "Dial 911" link on your dashboard, and until such later date, that such activation has been confirmed to you through a confirming email. You acknowledge and understand that you cannot dial 911 from this line unless and until you have received a confirming email.
>
> 2.5 Failure to Designate the Correct Physical Address When Activating 911 Dialing
>
> Failure to provide the current and correct physical address and location of your Vonage equipment by following the instructions from the "Dial 911" link on your dashboard will result in any 911 communication you may make being routed to the incorrect local emergency service provider.

This is a heavy-handed contract item, but what it means is that you have to use Vonage's prescribed, email-based activation routine to use its 911 call routing. Of course, I'm not a lawyer, and I can't provide an attorney's interpretation of this agreement, so contact Vonage if you're unsure about it. Other providers might handle 911 call routing similarly, so make sure you ask before you sign up if 911 is a highly important feature.

The best way to deal with this intimidating contract is to know firsthand whether your TSP has you set up for 911 calling, or be ready for an emergency in case it doesn't. That's what you're about to do.

The Problems with VoIP Emergency Dialing

With a traditional phone line, the power for the line and phone comes from a central power source at the phone company's exchange switch. This means that even during isolated power outages, you can still make and receive calls—including 911 calls. With VoIP, your electric company and in-house electrical circuits provide the power. If a circuit blows or the electrical supply fails, you won't be able to make any calls.

This would also be the case if your Internet connectivity failed or experienced a VoIP-prohibitive traffic jam. You wouldn't be able to make calls, or you might not be able to hear or be heard. Neither would be acceptable in an emergency calling situation, yet broadband VoIP TSPs can't prescribe a solution to this problem. This is because the TSP doesn't control the traffic between your VoIP device, your ISP, and the rest of the Net that provides the data transport between your VoIP device and the TSP. Unfortunately, there aren't many solutions to these issues.

Hack a Compromise Solution

In the event of an emergency, you're going to want to know you can pick up the phone and reach help quickly. You can do a few things to ensure this.

Keep a Plain Old Telephone Service (POTS) line for 911 calls. By keeping a traditional phone line hooked up, you ensure that you can reach 911 using "the old phone," and you provide a line that your VoIP ATA might be able to use for 911 dialing. Many VoIP ATAs and VoIP-integrated broadband routers, such as the Zoom X5V and V3 routers, allow you to connect a standard POTS line that 911 calls can be routed to in case of an emergency. Check with your VoIP TSP to see if it supports this kind of connection.

Program your VoIP device with speed dial to mimic 911. If you absolutely can't keep a POTS line around (or you prefer not to bear the expense of one merely for 911 dialing), you might be able to get your VoIP equipment to somewhat mimic the real thing:

- Program speed-dial buttons or key combinations on your IP phone or softphone that will auto-dial the local fire department or dispatch center via its regular, non-911 number. You should be able to obtain the local 10-digit phone number for the emergency dispatcher by contacting the administrative office of your local fire department. Ask them to give you the phone number of the line where 911 calls are answered. If you get lucky, the person you ask will know what VoIP is and will understand why you're asking, but don't count on it.

- If that's a dead end, you can program speed-dial buttons or key combinations (maybe even 9-1-1 itself) into your IP phone or softphone as a shortcut for calling a trusted neighbor or family member. This isn't exactly emergency dispatching, but it's better than nothing.

Use a cell phone for 911. Like a POTS line, a cell phone can often be used effectively to reach the 911 dispatcher, but check with your cell phone carrier to make sure 911 service is available and reliable in your service area. Just because wireless 911 service has been mandated by the Federal Communications Commission doesn't mean it works everywhere, so check with your carrier to be sure.

Use a good old-fashioned permanent marker. If all else fails, using a felt-tip permanent marker, write the full 10-digit phone number of the local public safety dispatcher on every phone in your house that uses your VoIP service. Don't write it on tape or sticky labels adhered to the phone, because they will eventually peel off, and you never know when you'll need that important phone number.

Update Your VoIP ATA Firmware

HACK #7

An ATA with up-to-date firmware will have fewer problems.

"Yesterday, I made phone calls through my VoIP TSP all day long! But today, I don't even hear a dial tone when I pick up the phone!" grumbled the frustrated consumer, regretting having replaced his local telephone service with a slickly advertised VoIP service from a California company called Ownage. This was the third or fourth time his VoIP service had quit working. So he grabbed his cell phone and frantically called the Support Department at Ownage.

The tech who answered wasn't especially helpful. She listened to the customer describe his recurring problem and then told him the same thing Support had been telling him ever since the first time the dial tone disappeared: "Sir, can you reboot your analog telephony adapter by removing the power cord and then plugging the power cord back in again after a few seconds? That should take care of the problem."

"But I do that every time. Ma'am, I bought this *VoIP Hacks* book that taught me how to wire my ATA into my home phone wiring so that I could replace my local phone service with Voice over IP, and now I'm very frustrated because every few weeks, I pick up the phone and the dial tone is gone. I have to run downstairs and reboot my ATA before I can place any calls, and I'm a little frustrated," the exasperated customer said. "Why is this happening?"

"Well, it's actually quite simple. The ATA receives an IP address from your DHCP server, which runs on your broadband router," she explained. "And your broadband router receives an address from your Internet provider's DHCP server. That IP address can change sometimes, when your DHCP lease expires, breaking the UDP socket that connects your ATA with our network here at Ownage."

"In English, please?" the customer said.

"Well, the problem occurs because your ISP assigns you a dynamic address that periodically changes," the support tech explained. "When it changes, the ATA loses communication with our VoIP server."

"So, it's my ISP's problem?"

"No, not exactly. Most ISPs use dynamic addresses for residential broadband customers to prevent them from, say, hosting their own servers. So, they have their reasons for using dynamic addresses, and there's little we can do about it," she told him.

"Then what do I do to stop it from happening again?" the customer asked.

"I'm glad you asked," she replied. "You can download the latest firmware patch for our ATA, which should make the ATA automatically reregister with our server whenever it loses communication. That would be the best thing to do."

"Is there anything else I can do?"

"If you'd like to hack a solution, you could build a system that can perform a regular, timed reboot of your ATA. Or you could cron a shell script that dials into a Dataprobe AutoPAL—this is a really cool device that lets you remotely reboot things—or you could...."

But he cut her off. "OK, I think I'll just download the firmware patch. Can I get it from your web site?" he asked.

"Of course. Is there anything else I can do for you today?"

"I don't think so," he said.

"Thanks for calling Ownage, sir. Have a good day," she said, and they both hung up. Satisfied, the customer picked up a pen and jotted down the entire conversation in the hopes of someday publishing it in a book about VoIP.

 You can find out more about the Dataprobe AutoPAL at *http://www.dataprobe.com/power/auto_pal.html.*

The Hack

Updating ATA firmware is a great way to stay on top of known perfor-mance issues—and it might allow you to take advantage of new telephony features introduced by your TSP. Most telephony hardware vendors tend to make their systems more stable with each release, so understanding your TSP's prescribed method for installing firmware patches onto your ATA is important. I've chosen Packet8 for this example. Refer to your specific TSP's support site for details on its update procedures.

Get the firmware update. Packet8, for one, offers a Windows executable that you can download from its web site (*http://web.packet8.net/download*). This tool will automatically identify your Packet8-provided DTA-310 ATA, download the patch, and install it. If you prefer not to use the tool, you can install the patch using the Packet8 ATA's web interface. Instead of down-loading the executable installer tool, just download the firmware file. Save it and remember the path where you saved it.

Locate your ATA. Next, if you don't know your ATA's IP address, use Packet8's IP-address identification service to find out what it is. This will be helpful if you've forgotten it, or if your ATA is configured to get its IP address via DHCP. Simply pick up your phone and dial 0120003. This will play back a recorded greeting that includes your ATA's IP address.

Next, go to that address with your web browser, using a URL like this: *http://10.1.1.200*, replacing *10.1.1.200* with your ATA's actual address.

When the ATA welcome page appears, click the Upgrade Firmware link. Click the Browse button to locate the firmware image file you downloaded earlier. Then, click Start Firmware Upgrade. After your ATA has rebooted, the update will be finished.

Desktop Telephony
Hacks 8–27

To take advantage of computerized telephony, you don't need a VoIP gateway, a fancy Internet Protocol (IP) phone, or an open source PBX (though those are certainly fun, hackworthy telephony goodies). Your desktop PC can be the nerve center of all your voice communications, replacing your telephone, your caller ID display, your answering machine or voicemail, and possibly even your phone bill (some VoIP services will bill you electronically).

Some pretty amazing software goodies are available to make your voice communication life a real joy. Programs like Gizmo Project and Skype let you make voice calls to buddies around the globe—for free. Some of these programs have built-in voicemail and call recording, and most are cross-platform, offering support for Mac, Windows, and Linux.

Hardware contraptions and telephony automation software bring even more exciting capabilities to the table. With a telephone-line interface for the Mac or a voice modem in a Windows PC, all you need is the right software to tie your phone completely to your desktop—but don't forget your wireless headset. VoIPing is much cooler when you aren't physically bound to your PC.

So don't delay; dig in to this grab bag of desktop telephony ideas. They're just the tip of the iceberg.

HACK #8   Access Next-Gen Voice Features

Broadband VoIP providers like Vonage don't just provide phone service. If you know where to find the features, they integrate with other applications on your desktop—and with your digital life.

When you subscribed to your amazing new VoIP telephone service, you might have missed the fact that, along with your new Internet calling, money-saving VoIP service, you also picked up some nifty desktop telephony enhancements. Most of the broadband VoIP phone service providers give you some cool extras that you'd never get with a traditional phone company—stuff like web-based account management, voicemail-to-email integration, and even softphone calling from your desktop. Did you know...?

Vonage Users Can Call Any Outlook Contact with One Click?

Vonage lets you place calls to your Outlook contacts with a special piece of software, an add-in called Click-2-Call, which comes on the Vonage software CD. Install it and launch Outlook. You'll notice that your Outlook contacts now have a Click-2-Call option in their Actions menu. Clicking this option dials the contact's phone number via your Vonage analog telephone adapter (ATA) and then connects the call with your phone. Pick it up; you should hear your call ringing in the handset, waiting for your contact to answer.

BroadVoice Users Can Use a Web-Based Tool to Place and Manipulate Calls?

If you're a BroadVoice subscriber, you've got some really cool web-based call-management tools at your disposal. Thanks to BroadVoice Call Manager, a web-based tool that BroadVoice gives you access to when you sign up, you can use a web page to control your voicemail, enable and disable call forwarding, and even tell BroadVoice how to handle your incoming calls based on their caller IDs—maybe you want to forward certain callers to one number, while allowing your BroadVoice voicemail to handle other callers. Nifty, eh?

You Can Automatically Dump Unwanted Girlfriends and Boyfriends Using a VoIP-Based Service?

Sad, but true. Hey, if you can get a date using the Web, why not dump people the Internet way, too? VoicePulse, a broadband VoIP carrier, provides the VoIP network framework for a service that will help you handle unwanted advances like a dating champ. You don't have to be a VoicePulse subscriber to use the service, though. Any phone user—VoIP, traditional, or cell—can dump somebody the high-tech way.

Let's say you're at a party and some doofus asks you for your phone number. Give the doofus the local number you find at RejectionHotline.com (*http://www.rejectionhotline.com/numbers_and_cities.php*), rather than your real number. When the dork calls for a date, he or she will instead get a professional rejection courtesy of the Rejection Hotline.

Aside from being cruelly entertaining, the Rejection Hotline provides a great demonstration of a large-scale soft-based voice system. By the time you're done with this book, you'll probably have enough VoIP chops to start your own version of the Rejection Hotline.

Broadvox Direct Users Can Use Find-Me-Follow-Me so that They Can Be Reached Wherever There's a Phone?

You bet! When you subscribe to the Broadvox Direct VoIP service, you get a web-based toolset that lets you configure a find-me-follow-me call list. That way, when folks call your home phone, the service can attempt to track you down on your cell phone, at Mom's house—wherever you might be.

HACK #9 Track Vonage Account Info on Your Desktop

This tiny desktop tool helps keep track of your minutes and voicemails, too.

If you've never used Konfabulator (now known as Yahoo! Widgets) or Apple's Dashboard widget system, you should try it out. *Widgets* are very simple, specialized desktop apps that provide short, useful information in real time. They can be floating windows, or they can be embedded into your desktop. Remember Active Desktop from 1997, which let you dock an informational web page into your Windows desktop? Well, widgets are about nine times better.

The widget experience is best with Yahoo! Widgets, a widget framework that seamlessly integrates with Mac OS X and Windows—specifically, Mac OS X 10.2 and higher, or Windows 2000 and XP. Literally thousands of different widgets are available that run on both Mac and Windows—everything from weather reports and stock tickers to cute little iPod remote controls and telephony-related goodies. One such goody is the must-have vonageGauge widget by Martin Koistinen, which gives you a one-glance update of your remaining Vonage minutes, as well as a count of voicemails waiting to be listened to (Figure 2-1).

Installing Yahoo! Widgets

It's quite worth your while to install Yahoo! Widgets, even if you can't benefit from vonageGauge. Throughout this book, I reference a number of other cool Yahoo! Widgets that will aide you in your telephony travails. The place to start is *http://widgets.yahoo.com/*. Here, you can download a version of the Y! Widgets system for either platform.

To install on Windows, just run the installer that you downloaded. To install on Mac OS X, drag the Konfabulator icon (which might eventually

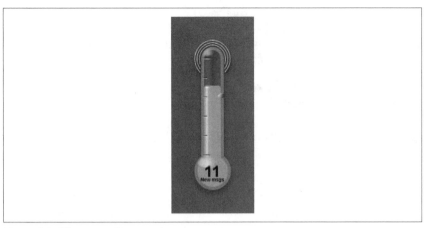

Figure 2-1. The vonageGauge widget in action

become the Y! Widgets icons) from the downloaded DMG volume folder into your system's Applications folder. Then, launch it by double-clicking it.

You'll be stepped through a wizard that helps you decide where you want to store downloaded widgets as your inevitable widget addiction grows. When the wizard is complete, Konfabulator/Yahoo! Widgets will automatically launch its default set of widgets. Now, to try out vonageGauge.

Installing the Vonage Widget

Download the vonageGauge widget from *http://www.widgetgallery.com/ view.php?widget=36334* and save it in a temporary folder or in the standard spot where you put downloaded files. Unzip the download (or mount it by double-clicking it, if on a Mac) and copy the enclosed widget file to the widget folder you selected during installation.

Mac users can launch the widget with no further issues. Windows users, however, must do some manual configuration due to lack of SSL support for the *curl* web utility in the Windows version of Konfabulator. Don't worry, though. This is hardly a painful thing to fix. You need this because Vonage's web site requires (as it should) SSL encryption to be employed when accessing account information.

To rectify the matter, download the most recent version of *curl* from its web site, *http://curl.haxx.se/latest.cgi?curl=win32-ssl*. From the downloaded zip file, note the files *curl.exe* and *curl-ca-bundle.crt*, as you'll need them in a moment.

Then grab the needed SSL libraries, *ssleay32.dll* and *libeay32.cll*, from *http:// www2.psy.uq.edu.au/~ftp/Crypto/*. The libraries will be located in a *binaries* directory on one of the FTP mirrors listed here. The specific file you need to download will have a name like *SSLeay-X.X.X-DSA.msw32.zip*. From inside this zip file, copy the two SSL *dll* files to the Konfabulator *wbin* folder located at *\Program Files\Pixoria\Konfabulator\UnixUtils\usr\local\wbin*.

Copy the two files from the *curl* download here, too. In total, you should have copied four files into this folder.

Gauging Your Vonage Utilization

Launch the widget by double-clicking its icon in your widgets folder. The thermometer-like display shows you how many minutes are remaining on your monthly plan. The more minutes you use, the lower the height of the "mercury" in the thermometer. This can help you conserve your utilization and spread your usage out to control your Vonage burn rate. Note that if you have an unlimited plan, this isn't really doing much for you, aside from showing you how many voicemails you have waiting.

To listen to unheard voicemails, double-click the text at the bottom of the thermometer, and your web browser will launch Vonage's services page, where you can hear them.

HACK #10 Pick a Desktop VoIP Client

There's no shortage of fantastic VoIP software for Windows, Mac, and Linux. But which one (or two) do you need?

VoIP applications tend, like email, to have a few servers facilitating interaction on behalf of many clients. In the case of email, those clients are applications like Microsoft Outlook, Eudora, and Apple Mail. But in Voice over IP, clients can be standalone devices, like IP phones and interface boxes (ATAs like those described in Chapter 1), or desktop applications like softphones or instant-messaging apps. The information in this hack will help you decide which VoIP client is right for you.

Some VoIP clients use well-known standards such as the Session Initiation Protocol (SIP) and are designed for use with your choice of VoIP service providers. Others are designed specifically to attach only to a certain service—such as AOL Instant Messenger (AIM). Still others are built using open standards but are hard-wired to work with only certain services; Yahoo! Messenger uses SIP but works only with the Yahoo! service. That is, you can't use the VoIP features of Yahoo! Messenger with your own choice of VoIP service providers.

Some VoIP clients are quite functional "out of the box," such as Skype, which provides a user-friendly wizard to sign you up for Skype service and get you logged in. With others, such as X-Lite and GnoPhone—which are designed for use with your choice of service providers, or even with your own VoIP server—you really need to know what you're doing to get much use out of them. Since X-Lite and GnoPhone aren't officially sanctioned for use with a particular provider, you've got to know *how* to configure them yourself.

Meet H.323, SIP, and IAX

VoIP clients and servers use three common standards for signaling call events. (These events might be the beginning and end of a call, an attempt to join a voice conference, or looking up a phone number to discover the best way to reach a particular user on a VoIP network.) These three communication protocols are H.323, SIP, and IAX. Very rarely does a single client support more than one of these protocols (Firefly is an exception, and provides support for both SIP and IAX). Having a basic grasp of the different protocols will help you choose a VoIP client.

H.323: the earliest VoIP standard. An H.323 client, such as Microsoft's Net-Meeting, really is good only in a corporate telephone system environment. It was once fashionable to use H.323 to have voice conversations with buddies over the Internet, but the rise of broadband firewall routers—which break the H.323 signaling protocol—and the growth of better protocols such as SIP led to a backslide in NetMeeting's popularity as a personal VoIP tool. Microsoft has since replaced much of the functionality of NetMeeting in its Windows Messenger IM software. So unless you need a softphone that works with your H.323-based PBX system (like an early-model Nortel PBX or Cisco media gateway), you're probably best served by foregoing H.323-based software.

GnomeMeeting is a very NetMeeting-like application for Linux.

SIP: the dominant VoIP standard. SIP has become the dominant multimedia communication protocol, used by an overwhelming majority of VoIP service providers and professional phone system vendors. Aside from voice, you can use SIP components to signal video and instant-messaging conversations, too. I'll concentrate on SIP as it applies to voice, though.

There are two kinds of SIP VoIP clients: those that allow you to connect to a VoIP system of your choice and those that are programmed for use only with a certain provider. SIP-supporting VoIP client software includes products such as Yahoo! Messenger, Apple iChat, sipXphone, Firefly, Gno-Phone, Gizmo Project, and lots of others.

IAX: a really cool VoIP protocol. Inter-Asterisk Exchange protocol (or IAX, pronounced *eex*) is used by a growing number of VoIP client programs and service providers. The coolest thing about IAX is that it's firewall-proof. In situations where SIP and H.323 are rendered inoperable by NAT firewalls like your home broadband router, IAX shines. The only problem is finding a service provider with which to use IAX (visit *http://www.teliax.com/* to learn about one that offers an IAX-based VoIP telephone service). IAXPhone and Firefly use IAX.

Understand VoIP Client Features

You ultimately will decide on a VoIP client based on features and compatibility. While one VoIP client might support the protocol you need—say, SIP—it might not support the features you need. iChat and X-Lite are both SIP software, but you can't use iChat with your own VoIP server; you need X-Lite for that. (If you're reading this book from front to back, you might be wondering if I'm planning to show you how to build a VoIP server. For the record, I *am*, but not until Chapter 4.)

Then again, the protocol or innards of the software might make absolutely no difference to you (plenty of folks use Skype, which doesn't use a standard protocol at all). Table 2-1 is a matrix of VoIP client software and their features and compatibility.

Table 2-1. VoIP client software compared

Software	Mac	Windows	Linux	Uses SIP	Uses H.323	Uses IAX	Use with your own server	License type
Gno-Phone	No	No	Yes	Yes	No	No	Yes	Open source
IAX-Phone	Yes	Yes	Yes	No	No	Yes	Yes	Open source
Skype	Yes	Yes	Yes	No	No	No	No	Freeware
sip-Xphone	No	Yes	No	Yes	No	No	Yes	Open source
AIM	Yes (no VoIP features)	Yes	No	No	No	No	No	Freeware
iChat	Yes	No	No	Yes	No	No	No	Free
X-Lite/X-PRO	Yes	Yes	No	Yes	No	No	Yes	Free/Comm

Table 2-1. VoIP client software compared (continued)

Software	Mac	Windows	Linux	Uses SIP	Uses H.323	Uses IAX	Use with your own server	License type
Firefly	No	Yes	No	Yes	No	Yes	Yes	Free
Gizmo Project	Yes	Yes	Yes	Yes	No	No	No	Free
Net-Meeting	No	Yes	No	No	Yes	No	Yes	Free
Gnome-Meeting	No	No	Yes	No	Yes	No	Yes	Open source

As you work through the hacks in this and the following chapters, you'll become very comfortable with the differences and similarities of these programs—and you'll have an even better feel for their strengths and weaknesses. A quick Google on any of these program names will get you to a place where you can download and install the program. And, speaking of Google, to get the most out of Google when using telephony, read "Google for Telephony Info" [Hack #21].

 ## Sound Like Darth Vader While You VoIP
#11

Using Audio Voice Cloak, you can sound like Darth Vader—or like Alvin and the Chipmunks—while you talk online.

Star Wars Episode III: Revenge of the Sith hit the screens right around the time I first tried this hack. When I filed into the very first midnight screening of the movie at my local cineplex, I was particularly excited by the prospect of again hearing the voice of the galaxy's most dysfunctional father. There's just something about James Earl Jones and the flange effect.

After all, who hasn't looked into a mirror in a private moment and said, "*I am your father!*" a few times? OK, maybe you're not as big a Star Wars geek as I am, but if you are a closet Wookiee lover, I've got the perfect hack for you to use the next time you chat with fellow fans.

> If you think spy movies are cooler than Star Wars movies, you can also use this hack to make yourself sound like one of those disguised-voice phone informants that sound a lot like, well, Darth Vader.

Gold Software's nifty voice-changing tool, Audio Voice Cloak, lets you tweak your speaking voice, adding pitch shifting, EQ, echo, and other sound effects in real time (Figure 2-2). If you have Windows, you're in luck (Mac folks, see the sidebar). Download and install AVC from *http://www.gold-software.com/downloads5903.html*. Launch it and, after the shareware commercial, you'll be able to click the All Controls button to reveal all of the sound-altering controls available to you. The program uses the default microphone input, so if you're using a nonstandard microphone channel for your telephony or online chat, you'll need to click the Recording Source button and select the right input.

Figure 2-2. Audio Voice Cloak's main interface

While you tinker with AVC's settings, you can monitor yourself with the aptly titled Monitor Your Voice button. Beware: you'd better put on a pair of headphones, or you'll get feedback.

To get the most authentic Vader imitation (short of hiring Ben Burtt, the famed sound effects guru from Lucasfilm), you'll want a slightly southerly pitch shift (drag the pitch slider down a notch or two) and a flange effect (click the Flange Off button to toggle it on). Finally, click the Center button on the Equalization panel to flatten (or "reset") the equalizer. Then, monitor your speech to hear how you sound. You should have the familiar, convincing tone of a half-machine Sith lord.

Now, fire up your Yahoo! chat client or AIM and surf on over to the closest chat room. Since AVC passes the modified audio through in real time, you can chat live as Darth, or you can raise the pitch shift to sound like a chipmunk. And don't discount the immaturity factor: if you have kids who chat with their buddies online, this could be a lot of fun!

Voice Alterations on a Mac

If you're a Mac user and you want to achieve the same voice alterations that Audio Voice Cloak makes possible for Windows users, you'll need to get your hands on a tool for Mac OS X called Soundflower. This awesome piece of software allows you to pipe audio into—and out of—applications in real time.

The "pipes" carrying the audio are logical OS X sound devices, so you can use them with any audio apps that support Core Audio, the standard sound framework on OS X. You can create a pipe to carry your raw audio into Pro Tools Free or Logic Express, run it through whatever real-time transformations you like, including pitch shift and flange for the Vader effect, and then send it out to your softphone or chat application using another pipe.

For more information on Soundflower, check out *http://www.macupdate.com/ info.php/id/14067.*

**HACK
#12**

Grow Your Social Network with Gizmo

If you love Skype but hate the fact that it isn't open and standards-based, you'll be right at home with Gizmo.

Gizmo Project, sponsored by SIPphone Inc. (*http://www.sipphone.com/*), seeks to create a free, peer-to-peer softphone with instant messaging à la Skype, but without the proprietary hindrances of Skype. In this regard, Gizmo does an excellent job. Its features are the same on Mac, Windows, and Linux, too—which means no more waiting two months for Windows-only features to show up in the Mac and Linux clients, something Skype users are accustomed to. Another cool plus that Gizmo brings to the table is free voicemail, something Skype has yet to offer.

To get started with Gizmo, hook up your headset and microphone, and download and install the client for your platform from *http://www. gizmoproject.com/*. Launch the Gizmo app, and register for your Gizmo name from the login screen. This name is both your login ID and the name that other Gizmo callers will use to call you. Once you're logged in, set up your user profile, as in Figure 2-3.

Figure 2-3. Gizmo's profile dialog

Don't forget to check the "List my profile in the White Pages for public searches" checkbox if you'd like to hear from other Gizmo users. Otherwise, they won't be able to find you when searching Gizmo Project's central user database. If you'd like to search for some buddies to add to your contact list, start by clicking the Search button in Gizmo's main window. Its search function, which is similar to but less elaborate than that of Skype, shows you the city, state, and country of each user, if they've entered that information in their profile. Gizmo also has a big selection of rather cool built-in avatars (buddy icons), or you can select your own image file to use.

Placing a voice call with Gizmo is as easy as entering the Gizmo name of the person you want to talk to and clicking the round phone icon in the upper right of the main interface window. If you don't yet have any buddies in your contact list, a great place to start is the Gizmo Project Party Line, which you can call by typing partyline in place of a normal Gizmo name. Calling the Party Line connects you to other folks in a voice chat room who might be able to help you start your social network with Gizmo. If there's nobody in the Party Line chat room at the outset of your call, you'll literally hear crickets chirping (how appropriate).

Extra Gizmo Features

Gizmo comes with a few extra features not available with a default install of Skype.

Map It. Have you ever wanted to know where the person you're talking to is located? When you're in a call, click the Map It icon, and you'll see a very nice satellite photo with lines drawn between the estimated locations of the call's participants, as in Figure 2-4.

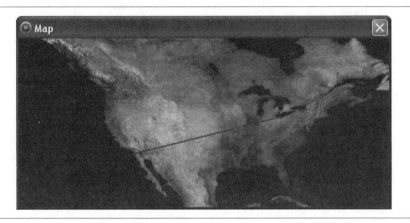

Figure 2-4. Gizmo's Map It function

Record It. Podcasters rejoice! Folks who've been looking for an easy way to record their VoIP calls from Skype and other softphone apps need look no further than Gizmo Project, which has the built-in ability to record all calls without the need for any other software. As a call is in progress, if you want to record it, just click the circular record button at the top of the conversation pop-out window that appears at the outset of each call. Recorded calls are saved in a WAV audio file on the desktop by default (you can change this location in Gizmo's preferences).

Gauge It. At the bottom of Gizmo's main window is an icon that looks like the signal-strength icon you might be familiar with from your cell phone—a row of vertical bars that indicate the quality of the connection to the phone network. In Gizmo's case, the bars represent the quality of the voice pathway your Internet connection provides. If you double-click the icon, you'll get a pop-up dialog that gives you more details about your available bandwidth, and you'll find out Gizmo's opinion of your Internet connection (apparently Gizmo doesn't particularly care for mine; see Figure 2-5).

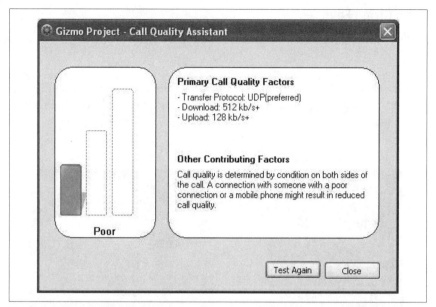

Figure 2-5. Gizmo tells you how well it expects to perform using your broadband connection

Share the Love

If you've come this far with Gizmo and still haven't placed an actual call, why not use that 25-cent call-out credit that SIPphone Inc. provides and call a buddy to tell him about it. Do this by typing your friend's phone number into the top drop-down list and clicking the round phone button. After a few seconds, your friend's phone will ring and you'll be able to talk for as long as a quarter will allow—which isn't long, so you might want to purchase more call-out credits by clicking the Out icon on Gizmo's Home tab.

Also Worth Checking Out

If you're really into desktop VoIP and you'd like to experiment with other standards-based messaging apps, I recommend JAJAH in addition to Gizmo. It's a Windows-based SIP softphone application that lets you call traditional phone numbers, like Gizmo, but also supports the IAX protocol and Skype, giving it the ability to communicate with several VoIP networks simultaneously.

I've always had to maintain several instant-messaging accounts to keep in touch with all of my online buddies on Yahoo!, AIM, and ICQ. To avoid running several instant-messaging clients, I have adopted Trillian, a Windows-based instant-messaging client that can talk to all of these networks, letting me manage all my IM activity from a single interface. (A similar Mac multinetwork IM tool is Adium.)

As Skype, Gizmo, and other VoIP networks grow, you'll probably need to attach to them simultaneously, as you would in the realm of IM. JAJAH lets you connect to VoIP networks using several major standards—SIP, IAX, and Skype—simultaneously, saving you from having to run several VoIP clients at the same time.

Record VoIP Calls on Your Windows PC

Unless you're using Gizmo, you probably can't record your VoIP calls without a little outside software assistance.

If you constantly forget things (which I do), or you're a private investigator (which I'm not), you might have wondered how to record calls so that you can listen to them later. Recording calls on traditional phones and IP phones is a simple matter of analog electronics (see Chapter 5), but recording soft-phone and instant-messenger voice calls is another matter entirely. Of course, you can set an old-fashioned tape recorder on your desk and press the Record button, but come on! In our digital world, there's got to be a better way, right?

Of course there is. You can find a handful of useful recorder apps at *http://www.download.com/* and *http://www.downloadsquad.com/* that can record WAV files and MP3s from any sound input or output on your Windows PC. One such application is Total Recorder, developed by High Criteria (Figure 2-6). In its default configuration, Total Recorder will record only the output (the person on the other end of the call), but not your voice.

To alleviate this, click Total Recorder's Recording Source and Parameters button and then check the "Record also input stream" checkbox. This way, your recording will be sure to contain both sides of the call. The "Remove silence" checkbox will enable a feature that doesn't save moments of silence into the recording. This might be useful if you record a ton of calls and review them regularly, as waiting through unneeded silence would certainly slow this process and use up more hard-disk space.

A real time-saver is found by checking "Convert using Recording Parameters specified below" and then clicking the Change button. In the dialog window that appears, you can adjust the sound resolution and the output format. Just about every sound codec you'd want is supported, from Windows Media to MP3. For even more sound-conversion goodness, be sure to check out "Create Telephony Sounds with SoX" [Hack #24].

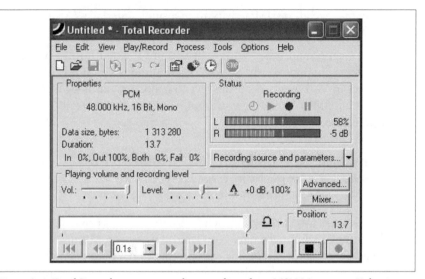

*Figure 2-6. Total Recorder can save audio recordings from MSN Messenger, Yahoo!
Messenger, AIM, Skype—you name it*

 ## Handle Calls with Windows Software

Have your PC screen your calls and take your messages with handy Windows
tools.

If you've got a Windows PC with a standard modem (it doesn't have to be a
voice modem), you can use some really cool software applications that can
identify incoming calls, with their caller ID information shown on your
Windows desktop. (Mac users can use Phlink for this purpose [Hack #15].)
Some of these apps can even respond to incoming calls so that you don't
have to.

This kind of application is a lot of fun, because while it technically doesn't
use any VoIP components (it's still strictly legacy phone technology), it will
give you an idea of how much power you as a phone user have when you
use software to enhance telephony applications. After all, your phone service is merely an application, and you'll be using a PC application to
enhance it.

> You'll need caller ID service enabled on your phone line if
> you want your PC to handle your calls in this way.

PhoneTray Free and PhoneTray Dialup

A cool freeware app that provides a caller ID pop-up window in the Windows system tray is appropriately named PhoneTray Free. In addition to the pop-up display, PhoneTray will log all incoming calls—handy when you've been out, and you want to know whom to call back—and it has a feature called Privacy Manager that lets you block calls from certain callers (Figure 2-7). PhoneTray also has a handy scheduler to establish your "quiet time," so you aren't receiving annoying calls in the middle of the night. While these features might be available from your local phone company, you can certainly save a few bucks by implementing them yourself with a PC tool like PhoneTray. The only hardware requirement is a modem connected to your phone line.

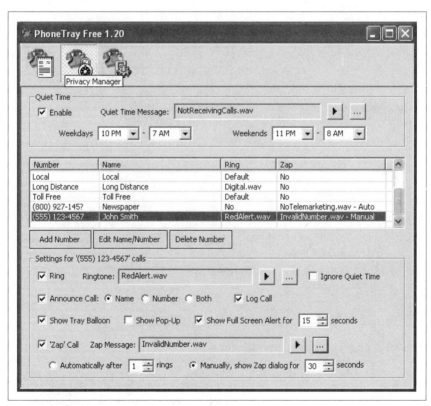

Figure 2-7. PhoneTray Free's Privacy Manager

For a small premium, PhoneTray's developer will sell you a version of the package, called PhoneTray Dialup, which works with caller ID–enabled modems. Using this feature, if you're a dial-up Internet user, you can receive caller ID signals on your PC desktop while remaining online. You can obtain PhoneTray from *http://www.traysoft.com/*.

Call411

Another nifty free application that handles caller IDs in your Windows workspace is Call411. While this app doesn't have the elaborate interface or the extensive call-handling options of PhoneTray, it's an effective, no-frills tool for displaying caller IDs, as shown in Figure 2-8. You can associate a custom ringtone with each caller ID if you like, and Call411 will even audibly announce incoming callers' phone numbers if you wish. You can obtain Call411 from *http://www.soft411.com/company/Soft411/Call411.htm*.

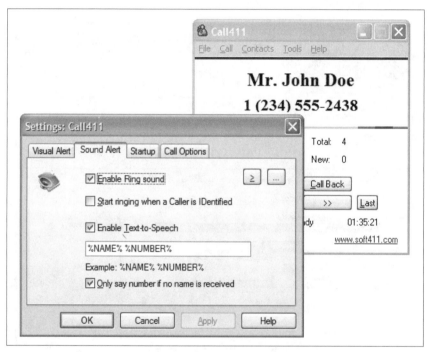

Figure 2-8. Call411 is a basic caller ID display tool

Tools like PhoneTray and Call411 are ideal for traditional phone service, but should work OK with VoIP services like Vonage and Packet8, too, since their ATA behaves just like a traditional phone line, with caller ID and all.

Call Soft and Call Soft Pro

If and when you outgrow Call411 and PhoneTray's features, and you find yourself wanting a complex voicemail and auto-attendant to handle your calls, you can graduate to a more capable commercial application such as Call Soft Pro, available from TOSC (*http://www.toscintl.com/*) and shown in Figure 2-9. This full-featured telephony system lets you tap into many of the standard calling features on your phone line. Even things like distinctive ring are supported. Call Soft Pro is a message recorder and interactive voice-response tool, so your callers can be prompted to select which voicemail box they'd like to record their message in. If you've got a large household, this can be a godsend. Mom and Dad can have the generic family voicemail greeting, and teenagers Todd and Susie can feel cool because they have their own individual voicemail greetings. With distinctive ring enabled on your phone line, the kids can even have their own separate phone number, and Call Soft Pro will recognize their ring pattern and play their special greeting.

> TOSC makes a scaled-down version of Call Soft Pro called Call Soft, which is geared more toward home users and lacks many of the automation features of the Pro version.

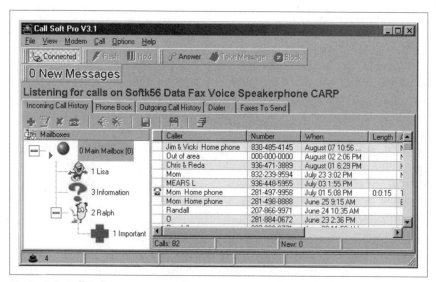

Figure 2-9. Call Soft Pro is a comprehensive message recorder and telephony package for Windows desktops

The program also offers a searchable call log, can receive faxes (which are saved as bitmapped files so that you can view or print them), and turns your PC into a speakerphone so that you can listen to calls through your PC's audio output and speak into your PC's microphone input. This allows you to use your PC as a phone and to record your conversations, instead of merely using it for automated call processing. Of course, there's a good bit of that in Call Soft Pro, too. Automatic forwarding of recorded messages via email is supported, and Call Soft Pro can provide music-on-hold for your callers, too.

Let Your Mac Answer and Log Your Calls

#15

Ovolab's amazingly simple Phlink telephony package lets you do some really cool stuff on your Mac—like answer calls and remotely control your Mac with a telephone call.

Watching Steve Jobs pitching the digital lifestyle at Macworld Expo is a favorite pastime of Mac enthusiasts. In fact, there's little that Mac users love more than watching the leader of the Mac world tout new developments and cool little tweaks in Apple's flagship iLife applications: iTunes, iPhoto, iDVD, iMovie, and GarageBand. But for all the pomp and circumstance surrounding these ravishing rollouts, Apple seems to have missed a critical component of the digital lifestyle, one that was around long before DVDs or MP3s—telephony.

Fortunately, an Italian company called Ovolab has created a really cool application that serves as the missing telephony link for iLife. Phlink is a hardware-software combination that answers calls with a voicemail greeting, logs them, and even allows you to set up AppleScripts that you can control remotely from a touch-tone phone. The hardware piece of the Phlink setup is a USB device with two RJ11-type ports, one for your standard phone line and another for your analog legacy telephone. The software component (available at *http://www.ovolab.com/phlink/*) consists of an application that looks like iTunes (see Figure 2-10). You get all of this for less than the cost of dinner (at a really nice restaurant).

Installing Phlink is a snap. Just plug the USB interface into an available spot on your Mac or its keyboard. There's no power adapter to worry about, thankfully. Plug your phone line into the "line" port on the USB interface, and plug your analog phone into the "aux" port. Then, drag the Phlink icon from the included CD-ROM to your Mac's Applications folder.

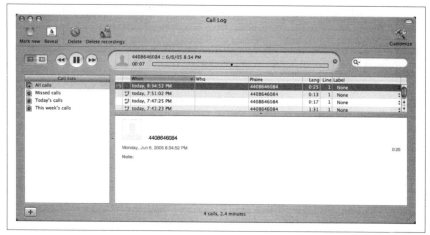

Figure 2-10. Phlink's main interface

Pop-Up Caller ID Notifications

In Phlink's Preferences window (available by clicking Preferences from the Phlink menu), you can enable an option that shows you a pop-up window with the caller IDs of incoming calls on your screen, so you can decide whether you want to answer them without having to even take your eyes off the screen, let alone leave your desk.

What's cooler than that? Well, how about telling your Mac to answer the call so that you can get on with what you're doing and not be bothered with answering the phone. To do so, just click on the round phone button in the Phlink action window, as shown in Figure 2-11. This starts the greeting to the caller and records the caller's message.

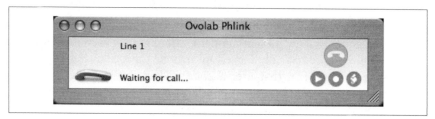

Figure 2-11. Phlink's action window

Custom Greetings

When Phlink answers each incoming call, it looks in the *Phlink Items* folder in the *Library/Application* support folder of your user profile for a file called *greeting.txt*. If it finds the file, it uses the Mac's built-in speech synthesis to speak the words in the file to the incoming caller. To modify this greeting, simply refill the contents of this file with whatever you like.

If Phlink can't find the *greeting.txt* file, it looks in the same directory for *greeting.aif* or *greeting.mp3*. Whichever of these files is present is played back to the caller. To create your own audio greeting, use a recording program (such as Cacophony [Hack #22]) to record and mix this greeting as you see fit.

> To get files into the right format for a greeting—i.e., AIF or MP3—use SoX. Refer to "Create Telephony Sounds with SoX" [Hack #24] for tips.

Answer Fax Calls

Phlink can answer fax calls. In the Preferences window on the Fax tab is a checkbox to enable automatic answering of incoming fax calls. You can have the Mac OS X fax viewer handle the faxes, or use the script option to handle them yourself. And speaking of scripting, now that you've covered the basics of Phlink, let's begin customizing!

Run Phlink Even When Logged Off

HACK #16

Phlink is a great application, but it's a desktop program, not a server app that's made to run in the background. So, when you log off, it shuts down and can't answer calls—unless you customize it to do so.

To get Phlink to launch upon login is really easy—just make it a login item for your user account in OS X Preferences. But getting Phlink to stay running even after you've logged out is a challenge. Of course, Phlink is most useful when it's running at all times, so you need to be able to do this.

Thankfully, for every unique need there's an equally unique hack. In this case, we're going to launch Phlink under a different user account. This user account will be automatically logged in at boot time, allowing the Phlink application to launch in that user account. Then, we'll create an AppleScript login item to switch to the user-selection screen automatically, giving you the option of logging in as any user you want.

To get started, open up System Preferences and click the Accounts preference pane. Click the + icon to create a new account, and call it Phlink. Make sure it is set to log in automatically upon startup. Now, log in as this user. Be sure to enable Phlink as one of its login items. To enable login items for the Phlink user, return to System Preferences and select the Login Items tab. Now you can add Phlink to and remove it from this list, causing it to launch whenever the user *phlink* logs in.

Now, launch the AppleScript editor, and create a script with this single line:

```
do shell script "/System/Library/CoreServices/Menu \\ Extras/User.menu/
Contents/Resources/CGSession - suspend"
```

The purpose of this one-line AppleScript is to present the user-switching dialog on the screen. We'll use this AppleScript to get back to the traditional login screen once the Phlink user has logged in and the Phlink app has launched. Save this AppleScript and then make it a login item for the user *phlink*. (Be sure it's listed *after* the Phlink login item.) Then, save and exit Preferences.

Now you're ready to try it. Reboot your Mac. If all goes well, your Mac will log in as the Phlink user automatically, launch Phlink, and return you to the screen where you can choose which user to log in as (or the username and password prompt, if that's how your Mac is configured).

Now, Phlink will handle your calls even while you're logged in as another user, or logged off altogether, because it's running as its own user in the background.

Greet Callers Differently Each Day

HACK
#17

Many voicemail systems let you use a different greeting depending on the day of the week or the time of day, but not Phlink—that is, unless you know how to use cron.

As you know from "Let Your Mac Answer and Log Your Calls" [Hack #15], a file in the */Library/Application Support/Phlink Items* directory contains your outbound greeting. Either it's *greeting.txt* (for a synthesized voice greeting), or it's an audio file in the form of *greeting.aif* or *greeting.mp3*. But suppose you want to use a different greeting depending on the day of the week.

Thank goodness for cron—the trusty Unix relic is a workhorse. Assuming you have all of your daily greetings stored in the same place, you can create a script that cron can use to update the greeting based on the day of the week. Here's the directory listing on my machine:

```
Mac-Mini:/Library/Application Support/Phlink Items kelly$ ls -al
total 328
drwxr-xr-x  10 tedwalli  admin     340    Jun  9 22:39 .
drwxrwxrwx  16 root      admin     544    Jun  6 19:50 ..
-rw-r--r--   1 tedwalli  admin    6148    Jun  6 19:50 .DS_Store
-rw-r--r--   1 tedwalli  admin      94    Jun  6 19:50 greeting.txt
-rw-r--r--   1 tedwalli  admin  755002    Jun  9 22:39 greeting_friday.aif
-rw-r--r--   1 tedwalli  admin  740222    Jun  9 22:39 greeting_monday.aif
-rw-r--r--   1 tedwalli  admin  700101    Jun  9 22:39 greeting_thursday.aif
-rw-r--r--   1 tedwalli  admin  694450    Jun  9 22:39 greeting_tuesday.aif
-rw-r--r--   1 tedwalli  admin  801006    Jun  9 22:39 greeting_wednesday.aif
-rw-r--r--   1 tedwalli  admin  154102    Jun  6 19:41 ring.aif
```

Now, create a quick shell script like this one, for each day of the week:

```
#!/bin/sh
## This script is for Monday.
cd "/Library/Application Support/Phlink Items"
cp -f greeting_monday.aif greeting.aif
```

Save each daily shell script in a convenient place, perhaps in a *cronjobs* folder in the user profile of that *phlink* user we made in "Run Phlink Even When Logged Off" **[Hack #16]**. Don't forget to make them executable (run **chmod 755 *** in the directory where you've put them). Then, add each file to the last column of the */etc/crontab* file, which defines scheduled Unix activities that should run in the background on Mac OS X. In this example, the daily file rotation occurs at 6:30 a.m. from Monday through Friday (the Friday greeting remains in place until Monday morning):

```
# cron jobs to rotate the phlink greeting
30   6    *    *    fri   ~phlink/cronjobs/Friday.sh
30   6    *    *    mon   ~phlink/cronjobs/Monday.sh
30   6    *    *    tue   ~phlink/cronjobs/Tuesday.sh
30   6    *    *    wed   ~phlink/cronjobs/Wednesday.sh
30   6    *    *    thu   ~phlink/cronjobs/Thursday.sh
```

Now, your Phlink setup will have a different greeting depending upon the day of the week.

Use Caller IDs in AppleScripts

#18   One of Phlink's AppleScript hooks occurs when incoming calls arrive, which means you can create actions to handle how those calls are handled.

If Phlink didn't have AppleScript support, it wouldn't be nearly as cool as it is. In fact, when I first fired up the Phlink application, I looked at the minimal interface and thought to myself, "Is that *it?*"

The fact is that Phlink's most awesome functionality is in its AppleScript object model. By using Phlink's functions in tandem with other AppleScript-aware applications, you can do some very cool telephony automation, from music-on-hold to greeting callers with the Mac's speech synthesis in interactive stages. Anything you can retrieve into an AppleScript variable from other Mac apps, you can pass into Phlink functions for interaction with callers. The only limit, then, is your imagination.*

---

* For an unrelated affirmation of the limitlessness of the human spirit, visit *http://www.zombo.com*.

As I got into setting up these voice AppleScripts, I was reminded of the interprocess communication goodness of the vintage Arexx scripting language on my old Amiga 4000 computer. I got to thinking—wouldn't it be cool to do some voice hacks on that 25 MHz classic? Then I realized that the Amiga's pokey 680x0-vintage processors don't even have enough processing power to encode and decode modern audio codecs! My hopes of splashing the cover of O'Reilly's *Make* magazine with a really tasty Amiga VoIP hack were dashed, and I returned to the 21st century realm of *VoIP Hacks*.

A great place to start building Phlink AppleScript hacks is with caller ID. When Phlink receives an incoming call, the first script Phlink calls is *ring*, which you'll create in the */Library/Application Support/Phlink Items* directory. The call doesn't have to be answered to execute this script; the line just needs to ring.

Now, while I'm not going to give you a full-blown explanation of AppleScript (O'Reilly's *AppleScript: The Definitive Guide* does a far better job than I could hope to, anyway), these examples should suffice to let you hack Phlink. Since we're starting out with the *ring* script, take a look at this example, which announces the caller ID of the call while it's ringing:

```
on incoming_call given callername:theName
        if the Name is ""
            return true
        else
            say ("You are receiving a call from " & theName) as string
            return false
        end if
    return callAgain
end incoming_call
```

It will probably take two rings before caller ID information is transmitted from the phone company (it tends to come between rings), but Phlink calls the *ring* script until it returns false, as in the previous example. An even cooler use of the *ring* script is to retrieve the caller's Address Book entry based on the caller ID signals received:

```
on incoming_call given callername:theName
        if the Name is ""
            return true
        else
            tell application "Address Book"
                set selectedPerson to (the name of the person \
                    whose name contains theName)
            end tell
        end if
    return callAgain
end incoming_call
```

Control iTunes from Phlink

HACK #19

If you have a ton of iTunes tracks just sitting there on your hard drive, why not put them to work in Phlink.

One of the coolest things about Phlink is its AppleScript abilities. Much like PhoneValet and other desktop telephony packages, custom scripting is where all the fun lies. Sure, letting folks record their voicemail onto your computer is fun, but integrating the other stuff on your computer with the phone—that's even better. You've seen how to do some basic database interaction between Phlink and the Address Book [Hack #18]. That's a great starting point for this hack, because it introduces the events that can trigger scripts within Phlink. If you haven't been there already, check it out, come back, and I'll be waiting here with this iTunes hack.

Your iTunes music library makes the perfect source for on-hold music, or just for a cool telephone gimmick like a "remote phone jukebox." The following AppleScript will actually search through your iTunes music library and find non-copy-protected songs (i.e., songs imported from CDs or MP3 files, and not purchased online) to play for the caller:

```
on do_action given call:my_call
    tell application "iTunes"
    set track_found to false
    set num_retries to 0
    repeat until track_found
        set my_track to some file track of library playlist 1
        if (kind of my_track contains "Protected") is false then
            set track_found to true
        else
            set num_retries to (num_retries + 1)
            if num_retries > 100 then
                exit repeat
            end if
        end if
    end repeat
    set my_song_file to the location of my_track
    end tell
    tell application "Ovolab Phlink"
        tell the_call to play (my_song_file as alias)
    end tell
end do_action
```

my_song_file is a variable that stores the location of a song to play for the caller, which is triggered to play in the fourth-from-last line in the script. You can trigger this bit of AppleScript from any of Phlink's event-handling scripts (*ring*, *greeting*, *hangup*, etc.). The Ovolab Phlink user manual, written by fellow O'Reilly author Matt Neuberg, provides a scholarly introduction to all of Phlink's event-handling scripts.

Automatically Pause iTunes, Resume iTunes

Ovolab provides the following script to pause iTunes music playback when the phone rings. (This script is really cool. As Forrest Gump would say, that's about all I have to say about that.) Save it as *ring.scpt* (or modify your existing *ring*) and put it in the *Phlink Items* directory:

```
on incoming_call given call:the_call
    set my_paused to false
        tell application "Finder"
        if (exists of (every application process whose creator type \
        is "hook")) is true then
            tell application "iTunes"
            if player state is playing then
                set my_paused to true
                pause
            else
                set my_paused to false
            end if
            end tell
        end if
        end tell
    tell application "Ovolab Phlink"
        tell the_call
            make new bag with properties \
            {name:"pauseitunes", waspaused:my_paused}
        end tell
    end tell
    return false
end incoming_call
```

Now, to resume iTunes automatically when the phone call is hung up, add this to your *hangup* script:

```
on do_action given call:the_call
    tell application "Ovolab Phlink"
    tell the_call
    try
    if waspaused of bag "pauseitunes" is true then
        tell application "Finder"
        if (exists of (every application process whose \
        creator type is "hook")) is true then
            tell application "iTunes" to play
        end if
        end tell
    end if
    end try
    end tell
    end tell
end do_action
```

For more fantastic Phlink scripting magic, be sure to visit the message board at *http://www.ovolab.com/*, and for some more cool ideas for iTunes/Phlink scripting, visit the source of several of these scripts, *http://www.gunsmoke. com/scot/home_automation/phlink.html*.

HACK #20 VoIP While Fragging

This sure beats typing "OWNED!" in an in-game chat window.

If you're like most *übergeeks* (and I say this as an admitted *übergeek*), there might be no pastime more satisfying to you than online gaming. Indeed, it's hard to beat the pure excitement of fragging your best friend with a rocket launcher in *Quake* or laying down the Horde smack onto a *World of Warcraft* n00b. Of course, if you're a Ventrilo or Teamspeak user, you can use Voice over IP to rub it in your opponent's face verbally when you crush him.

Ventrilo and Teamspeak provide hands-free conference calling designed for online gaming. This way, teammates can coordinate their strategies verbally, communicating by mouth without interrupting their in-game action, rather than by typed messages, which can be a real distraction. Nothing's a greater mood killer than having to stop to type a chat message to call for a rescue, only to get hit from behind by a stray rocket while typing your plea.

One great feature of both Ventrilo and Teamspeak is their "push-to-talk" capability. This allows you to treat them like a walkie-talkie—cutting out the background noise that would otherwise be transmitted if the chat were always live. With this feature, you can even forego headphones if you keep your transmissions brief so as to discourage echo.

Ventrilo

Ventrilo, from Flagship Industries (*http://www.ventrilo.com/*), is a team voice chat system that uses the Global System for Mobile (GSM) codec—a very bandwidth-conservative codec that's excellent for use with games (you don't want your voice traffic to create in-game lag, so a codec like GSM is perfect). Ventrilo has client and server components. The client runs on Windows and Mac OS X, and the server runs on Mac OS X, Windows, and Linux. A version of the client for Linux is said to be in development.

To run the Windows client, you'll need DirectX 8.1 or later (available from Microsoft and standard with Windows XP and above). The Mac client requires OS X Version 10.3.2 or higher. You'll also need a microphone and a pair of headphones (the headphones are superior to using freestanding speakers, because ambient noise from the speakers will "spill" into the microphone, creating really annoying echo for your game-playing buddies).

Teamspeak

Teamspeak (*http://www.goteamspeak.com/*) is similar in purpose to Ventrilo, though its web-based chat room administration tools are more advanced, and its bent toward gaming is a lot more obvious (Ventrilo professes to be useful for other things in addition to gaming). Teamspeak uses the Speex codec, which, like GSM, is very lean on bandwidth, making it a good choice for lag-sensitive gamers.

Teamspeak offers clients and servers for Windows and Linux. They're fully interoperable with each other. Like Ventrilo, Teamspeak requires DirectX 8.1 (or later), and its designers insist on a pair of headphones to reduce echo. Also offered is a hosted, pay-for-play service based on the software.

Though not officially sanctioned by the designers of Teamspeak, a great Mac client called Teamspeex has been developed. You can download it from *http://www.savvy.nl/blog/download*.

Skype

Perhaps the easiest way to VoIP while fragging is with Skype. This desktop voice chat package supports conference calls with five participants—perfect for maintaining open communications for a marauding patrol of Halo warriors behind enemy lines. Skype has several things going for it: it has hands-free operation, making it ideal for gaming; it's fully cross-platform (even Linux has a client); and it's stable.

But since only five conference participants at a time are permitted in Skype (four plus the person hosting the conference), you aren't going to accomplish your entire 500-person *EverQuest* guild meetings using Skype. That kind of scalability is something you'd need Teamspeak for. For small conferences, Skype is adequate. Perhaps it's no coincidence that the maximum size of a quest group in *World of Warcraft* is five members—ideal for a Skype conference. Just be sure the host PC has plenty of horsepower and a solid broadband Internet connection.

Skype for Windows runs on Windows 98 and up, while Skype for Mac runs on Mac OS X 10.2 and up. For more Skype details, sift through the delicious goodies in Chapter 3.

The Skype Alternatives

Gizmo Project, which is a lot like Skype but uses industry-standard SIP for call signaling, is (as of Version 1.0) very limited in terms of conferencing. In fact, it doesn't have any VoIP-based conferencing built in at all. Conference calling using SIP requires a centralized conference-mixing server, a complexity that makes Skype preferable to Gizmo for in-game conferencing.

Google Talk (*http://talk.google.com/*) is another Skype alternative. Like Gizmo Project, Google Talk, which has two-party voice calling features, supports a well-known standard for call signaling, called Jabber. And like Gizmo Project, Google Talk is free. iChat can be configured for use with the Google Talk Jabber network, as can various other IM clients such as Trillian and Adium. So you aren't confined to using Google's Windows-only client if you want into the network. But that's where the pros end and the cons begin. Like Gizmo, there's no way to do conference calls. And worse still, voice chat between official Google Talk clients and non-Google clients such as iChat doesn't work at all. So Google Talk's usefulness as an in-game voice conferencing tool is, well, nonexistent.

The Hardware

To have voice communication while blasting the competition to bits, you're going to need headphones and a microphone. The mic can be built into the headset, or it can be freestanding. But, definitely use headphones. They'll cancel the acoustic feedback you would get if you were using regular speakers, and they'll substantially reduce annoying echo. Your gamer buddies will thank you. There's nothing quite as unnerving as a VoIP-enabled four-player round of *Warcraft III* when one of the players is echoing like crazy. Friends don't let friends frag without headphones.

H A C K   Google for Telephony Info
#21   Harness the world's most knowledgeable search database for your own voice purposes.

Near the end of the dotcom boom, a little search engine startup called Google was born. Today Google dominates search on the Internet. Though Google has moved into the realm of VoIP with Google Talk, its new IM client, the company's best offering to telephony is still its famously useful search engine, Google.com.

Mine for Phone Numbers

If you're looking for a particular phone number, or for a group of phone numbers to be used in telemarketing or fundraising applications, a great place to start is with Google. (And a really smart next step is the National Do Not Call Registry [*http://www.donotcall.gov/*] if you plan to solicit the folks you're calling. Once a call recipient informs you that his number is on the list, it's illegal for your organization to call him again.) Here are some Google search queries that you can use to turn up phone numbers.

Suppose you want to turn up numbers in a given area code and prefix. You can form your Google query like so:

```
"(440) 328" OR "440-328"
```

The quotation marks surrounding the two expressions tell Google to treat them each literally—that is, to return only instances of the entire expression ("*(440) 328*") and not mere instances of the elements within the expression (*(440)* or *328*).

Google will return web page hits that contain occurrences of the area code 440 and the prefix 328 (you might get some non-telephone-related stuff, too) in its two most common forms: with parentheses, and with a hyphen. Of course, the results you get from that query might require a lot of interpretation and massaging before you can really use the phone numbers that you've turned up in an automated dialing app or something similar.

Complete That Phone Number

Sometimes the results from a Google phone number search can be fast, useful, and simple to interpret. Let's say you need to call somebody in your neighborhood, like the local pizza parlor. Let's also say that you know the pizza shop's phone number begins with the area code that's common in your neighborhood. Bang a query like this into Google and you'll have the whole phone number—and your pizza—in no time:

```
Hungry Howie's Lakewood 216
```

By giving Google the name of the pizza place, the city it's in, and the area code you expect its phone number to have, the first Google result is (almost) always the right one—and the entire phone number you're looking for usually shows up in the short synopsis on Google's results page, so you don't even need another click. Hey, when you're craving pizza, time is of the essence, right?

Telephone Privacy Check

While you're perusing Google's phone number department, you might want to see if your phone number is in the Google search index. If it is, your privacy could be in question. Try Googling your phone number, with and without punctuation, and see what results come back. You might even turn up somebody who has published your phone number without your authorization. This isn't entirely likely, but I remember a couple of years ago when a list of thousands of valid credit card numbers made it into the Google index, so it's not out of the question.

Research VoIP History on Google Groups

Probably the best historical newsgroup search tool on the Web, Google Groups lets you go back in time to search for public correspondence about all kinds of topics, including Voice over IP. By surfing to Google Groups (*http://groups.google.com/*), I was able to find that the first mention of VoIP on record occurred in early 1996 and that the first mention of Voice over IP dates back to early 1995. A good deal of early IP telephony research is probably quoted in the Google Groups archives, so if you're interested in the history of VoIP, this is a fantastic source. After all, as Gavin DeGraw sings, "Part of knowing where I'm going is knowing where I came from."

Telephonize a Sound File

HACK #22

This trick is useful for taking on-hold music for a test drive, or just for making recordings sound like they're coming through a telephone.

Whenever you work with telephony, be it desktop telephony apps or full-fledged IP phone systems, you're bound to encounter prerecorded sounds; things like on-hold messages, voicemail greetings, and even elevator music are often generated by computerized telephony applications. You might even need to create sounds that can be used with these apps. Generating your own telephony-ready sounds is a snap using desktop recording software. You can even resample your recordings so that you'll know exactly how they'll sound in a VoIP environment. This way, you can "preview" them.

Sound-effects producers who need to make somebody's voice sound as though it's being heard through a telephone employ a technique called *downsampling*. This gives recordings that tinny telephone flavor. For a perfect phone-sound simulation, you'd also need to chop the high and low frequencies of the sound using an equalizer tool, but downsampling alone produces a pretty convincing "phone sound." Here's how it's done.

The easiest way to downsample a sound is by using a simple sound editing tool such as Richard Bannister's Cacophony for Mac OS X, or Windows Sound Recorder, which comes with Windows. In essence, you open the sound file, change its sampling resolution to 8 bits per sample and 8,000 samples per second, and then save the file. (On a standard telephone call, there are 8 bits per sample and 8,000 samples per second in the media stream.) This matches your prerecorded sound to the sampling resolution of a typical phone call.

In Figure 2-12, an MP3 sound file has been opened in Cacophony. Its left and right waveforms are displayed, since it's a stereo sound.

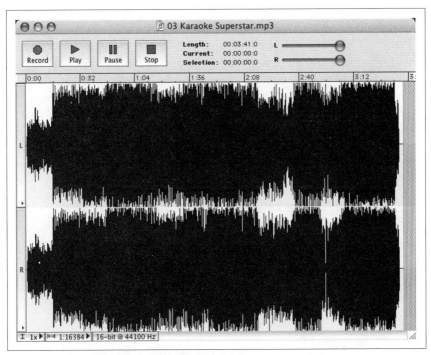

Figure 2-12. *An MP3 file opened in Cacophony—ready to "telephonize"*

Clicking Cacophony's Resample function displays the Resample dialog, which lets you specify the exact factors you'll use to downsample the sound. All telephony applications are mono, not stereo. All legacy telephone equipment (and most VoIP equipment, too) uses a sampling resolution of 8 bits and 8,000 Hz, as shown in Figure 2-13. Once these settings are made, clicking OK dismisses the Resample dialog and performs the downsampling on the sound file.

In Figure 2-14, the resulting sound waveform is telephonized. Listen to it now and hear the difference. It sounds much flatter, less crisp, and possibly more robotic. It sounds like you're hearing it through a telephone, which is the idea here.

 If you are considering putting on-hold music or background music on your VoIP system, this technique will let you hear roughly how it's going to sound. Some musical recordings might sound very poor once they've been downsampled, but now you'll know how they sound before your callers do.

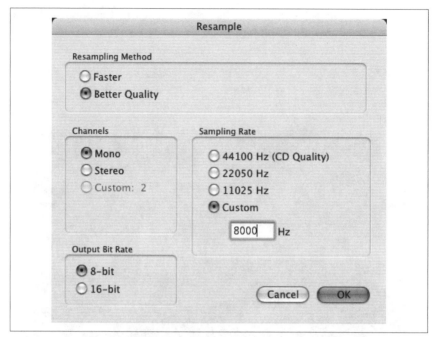

Figure 2-13. Cacophony's Resample function lets you downsample sounds from hi-fi to "tele-fi"

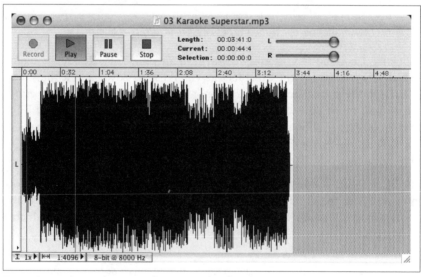

Figure 2-14. The resulting waveform is a mono, low-fi sound

Record an Audio Chat on Your Mac

Have you ever wished you had a recording of a past conversation?

Recording incoming audio from any application is a snap on Mac OS X, as long as you have Ambrosia Software's free trial version of WireTap Pro or Rogue Amoeba's Audio Hijack—two applications that permit you to siphon audio produced by desktop applications into sound files in real time. In this case, we'll use the X-Lite softphone [Hack #4] and Wiretap 1.0 to make an AIFF sound file that contains an incoming caller's side of the conversation. With this hack, you can record an audio chat on your Mac.

> To make a recording that mixes both ends of the conversation—your voice *and* the caller's voice—you'll need more elaborate sound-recording software, like the pay-for version of Ambrosia's WireTap (WireTap Pro), or Rogue Amoeba's Audio Hijack Pro. Both applications allow you to record from the Mac's audio output (for grabbing the caller's voice) and its audio input (for grabbing your voice).

First, we'll need to get WireTap (*http://www.ambrosiasw.com/utilities/wiretap/*) set up to make the recording. Since the audio fidelity of a phone call on the X-Lite softphone isn't likely to be higher than 8-bit/8 kHz, we'll configure WireTap to save using the same fidelity. Once you've launched WireTap, go to Preferences from its application menu and note the sound-compression settings. By default, the fidelity will be 44.1 kHz stereo and 16-bit sampling depth, as shown in Figure 2-15. Click the Settings button to change that.

As shown in Figure 2-16, you can drop the sample rate and depth to match an appropriate level for a telephone call (and consequently make your AIFF recording smaller). There's no point in saving 16-bit recordings if the audio coming from the recorded call is only 8 bits deep, and there's no point in creating a stereo recording of a phone call. Once you've made the changes, click OK to dismiss the Sound Settings window, and close the Preferences window, too, if you wish.

Next, fire up your X-Lite softphone (this will also work with Skype and iChat), and make sure it is registered to send and receive calls with your VoIP provider, as described in "Use a Softphone with a VoIP TSP" [Hack #4]. Then, call somebody or have somebody call you. As the call begins, click the red circle (record) button in WireTap's floating window, shown in Figure 2-17. This will start an AIFF recording of the current audio output, including the audio coming from the softphone. When the call concludes, click the square (stop) button.

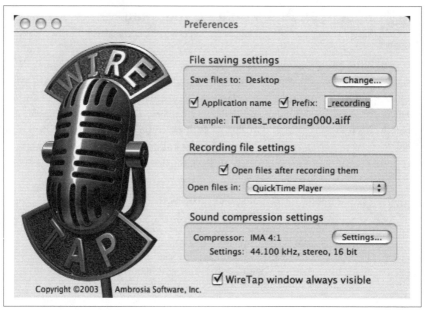

Figure 2-15. WireTap's Preferences window

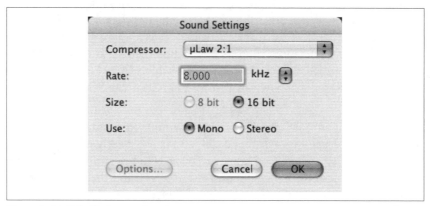

Figure 2-16. WireTap's sound settings

Figure 2-17. WireTap's recording controls; pretty simple

If you didn't disable the option to launch the recording automatically upon completion, it should immediately appear in a QuickTime window for you to listen to.

Create Telephony Sounds with SoX

Use the Swiss Army knife of sound-conversion utilities for your VoIP setup.

Though dozens of utilities are available for converting and tweaking audio files, the cross-platform open source audio tool called SoX really stands out. If you've got a Linux or BSD PC, chances are pretty good that you've got SoX installed. Windows and Mac users will have to download a compatible version of SoX from *http://sox.sourceforge.net/*. And since SoX is a command-line utility, you'll need to be at least a mediocre typist to get through this hack and the next two hacks. You'll also need to know how to *get* to a command line on your particular platform. On Windows, this means running the MS-DOS prompt. On the Mac, it's the Terminal. Linux and BSD users need only to fire up xterm. This hack will show you the ins and outs of using SoX to convert audio files from one format to another, add audio effects, and telephonize your audio through downsampling.

File Format Conversion

File format conversion is perhaps SoX's biggest strength. You can use SoX to convert from one format to another (WAV, AIFF, etc.) and from one encoding to another (uLaw, MP3, etc.). It even supports some fossilized sound formats like 8SVX and *.voc*. All of this format support is helpful if you want to use a file that you have only in some oddball format that your telephony software can't use.

In most telephony applications, like voicemail and interactive voice response (IVR), where recorded voice prompts are the user interface, you'll encounter sounds in one of a few encoding formats:

GSM
> An encoding commonly used on cell networks and in Voice over IP calls. This is the de facto encoding used by Asterisk voicemail greetings and other announcements.

uLaw or aLaw
> The two most common 8-bit pulse code modulation formats; they are most frequently used in legacy telephony, such as T1 voice connections and digital PBX telephone systems. uLaw is common in North America, and aLaw is common in Europe.

To convert a sound file from one format to another, there are two ways to go. SoX can recognize the input and desired output formats merely by parsing the filename extensions you provide, as in the following example:

```
$ sox basic_instructions.ulaw basic_instructions.gsm
```

This syntax takes *basic_instructions.ulaw* and creates a GSM-encoded file called *basic_instructions.gsm*. Of course, if the file you're converting doesn't have a file extension in its name, you can express your intentions more explicitly:

```
$ sox -t gsm another_brick -t aif another_brick_in_aiff_format
```

By specifying the encoding type with the -t option, you can tell SoX specifically how to convert the file, regardless of filenames and extensions. But that's not all you'll find SoX useful for.

Adding Sound Effects

Aside from converting files between formats, SoX can add some cool effects, too. Equalization, reversal, chorus, reverb, time shifting, and vibrato are some of the most commonly used effects options. Some of these effects are probably more useful in a pro audio environment than in VoIP, but there are uses for audio effects even in telephony—like an on-hold message that hypes a particular product or event. Such an announcement might benefit from a little reverb or delay. Just think about some of the sound effects used in monster-truck advertisements beckoning your attendance on *SundaySunday-Sunday!* Consider the following syntax, which adds reverb to a sound:

```
$ sox bigFootSunday.aif bigFootSundayVerb.aif reverb 1 1000 15
```

This example takes *bigFootSunday.aif* and adds 1,000 milliseconds of reverb with a 15-millisecond delay before saving the file as *bigFootSundayVerb.aif*. You can combine sound effects, too. So, for instance, you can place a reverb and an EQ effect together:

```
$ sox gilmour.aif gilmour.aif reverb 1 1200 30 highpass 1000
```

The reverb effect is followed by a high-pass filter, which is an EQ technique that trims (reduces) samples below a certain frequency—in this case, 1 kHz. You can experiment with the high-pass and low-pass features to trim frequencies, letting you obtain a number of cool effects. Make your music sound like it's coming through a megaphone or, with a little reverb, make it sound like you're singing in the shower. Now, it's up to you to find an appropriate venue for all this aural awesomeness in your VoIP setup—perhaps in your Skype answering machine [Hack #37].

Resample and Re-Level Sounds

The SoX bag of tricks has many compartments. Aside from EQ, effects, and format conversion, you can use SoX to downsample sounds, as in the Cacophony example earlier in this chapter [Hack #22]. SoX can also alter a sound's volume level (amplitude level).

To alter the sample rate, use the -r option and specify the desired sample rate in kilohertz. Of course, you can decrease (downsample) or increase the sample rate, but increasing the sample rate won't result in a higher-fidelity sound. This example takes a file called *bytor.wav* and downsamples it to 8 kHz:

```
$ sox bytor.wav -r 8000 bytor_8khz.wav
```

To alter a sound's amplitude, or volume level, use the -v option. This example increases the volume of the sound by 25% (using a negative value will decrease the volume):

```
$ sox -v 0.25 bytor.wav bytor_loud.wav
```

HACK #25   Mix the Perfect Announcement

Put SoX to work mixing different sound files—music, spoken words—to make the ultimate announcement message.

Are you looking for an easy way to make a seriously cool announcement for your Skype answering machine or outgoing voicemail greeting on your Asterisk VoIP server? You could buy a copy of a high-end audio package like Logic Express for all its cool sound-mixing and effects abilities. Even some of the simpler sound editing tools let you merge files, but (if you have Unix) chances are good that you've already got SoX! If you're on a Mac or Windows box, you're only a download away from having it, so take the MacGyver approach and save some cash. (Mac, Windows, and Linux users can get SoX from *http://sox.sourceforge.net/*.)

In this little project, we'll mix an announcement message with some background music, and then trim the resulting file to just the right length, all using the SoX toolset. Finally, we'll save it in the appropriate sound format.

To get started, find a piece of music that you think will make good background music—preferably something that has no lyrics to interfere with the spoken message you'll be mixing in. The music can be in any format SoX can handle—*.wav*, *.mp3*, whatever. If you can, note the length (in minutes and seconds) of the music file, as this could come in handy later. To figure out the length, just launch the file in your favorite sound player. It should show you the length.

If you use copyrighted music for commercial or nonprivate use, you'll need permission from the artist who created the music, or you'll have to pay royalties for using the music!

Next, record your announcement using Windows Sound Recorder, Cacophony, or your favorite sound recorder. Drop the resulting file in the same folder with your music file. Oh—and you might want to note its length, too. Use these two commands to perform some conversions on your raw audio files:

```
$ sox bg_music.mp3 -r 14400 -c 1 bg_music.wav
$ sox announce.wav -r 14400 -c 1 announce.wav
```

These two commands grab the files, resample them both to 14,400 kHz, and make them mono (-c 1), not stereo sounds, ready to be mixed together. The following command mixes the two resultant files into a single file:

```
$ soxmix bg_music.wav announce.wav mixed.wav
```

If you find that the background music is too loud or soft, you can adjust it and remix the files:

```
$ sox bg_music.wav -v -0.25 bg_music_quiet.wav
$ soxmix bg_music_quiet.wav announce.wav mixed.wav
```

Then, all that's left is to get the file into the format you need for your telephony application. If it's for an Asterisk announcement, you'll probably want it in GSM format at 8,000 kHz:

```
$ sox mixed.wav -r 8000 mixed_for_asterisk.gsm
```

Converting to other formats as opposed to GSM is as simple as changing the extension on your final filename.

HACK #26 Sound Like a Pro Announcer

Do you want to have a deep, commercial-sounding voice on your greeting messages? Here's how.

You might think it takes a ton of natural talent to sound like one of those professional, deep-voiced commercial announcers that you hear in the on-hold messages of big companies. But all it really takes is adherence to a few simple guidelines for clear speech…and maybe a little hacking talent.

First, the speech guidelines. Commercial announcers usually speak much more slowly and concisely than would be appropriate for normal conversation. As a result, they are very easy to understand and follow. Radio announcers often place almost lopsided stress on the words they want to emphasize, too. So, if your recorded announcements are sales pitches, you'll want to speak concisely, slightly more slowly than normal, and with great stress on the words you'd like to emphasize. You'll also want to smile while you record yourself. It seems odd at first, but a smiling announcer conveys a different attitude than a blank-faced one. A high school or college speech textbook might be a good source if you really want to hack your speaking ability!

Now, on to the hack. By now, you've probably used SoX to convert files from one format to another, but did you know SoX can deepen your voice like a pro announcer? To do this, you'll need to use SoX's shift (pitch) effect, which takes a positive or negative integer as an argument. If positive, the integer will increase the pitch of the sound by the number of steps specified. The higher the number, the higher the pitch will be. The lower the number, the lower the pitch will be. In this case, we're going for pro announcer, not chipmunk, so start with a value of -2 and work your way down until it's deep, but not artificial-sounding:

```
$ sox announce.gsm announce_lower.gsm pitch -2
```

HACK #27 Record a Videoconference

Snapz Pro X lets you record video and audio together—the perfect way to record a video conference.

The term *VoIP* usually refers to Voice over IP, but it could easily mean Video over IP, too, since video conferencing is such a popular use for the Internet today. Tools such as Yahoo! Messenger and Apple iChat allow you to do face-to-face video conferences across the Internet, but one thing neither of these tools allows is recording your conferences. Fortunately, Ambrosia Software's Snapz Pro X lets you create a QuickTime file of any onscreen activities, including a video conference. Unfortunately, it runs only on Mac OS X, so Windows and Linux users are out of luck.

You can download a copy of Snapz Pro X from *http://www.ambrosiasw.com/*. It will install as a background application that you can summon with a special key combination that you'll assign right after it installs, the first time it runs. When Snapz's main window appears, click the Movie button. Here, you can configure the size and aspect of the capture, as well as the frame rate that will be used in the saved video file and whether to include sound from the audio output (the conference participants' voices) or microphone input (your voice).

You can crop, resize, and drag Snapz's viewfinder so that it wraps tightly around the area of the screen you want to record. Just don't drag the window you're recording from out of the area of this viewfinder, as it remains in a fixed size, shape, and location throughout the recording (unless you've configured it to follow your mouse movements, which is probably not a good idea when recording from a video chat window). Once you've got everything positioned just right, press the Return (Enter) key, and Snapz will prompt you for a filename. Enter this and click OK, and recording will begin and will continue until you summon Snapz again using the key combination you established during installation.

Unless you change Snapz's default settings, the saved QuickTime file will appear in the Preview immediately after the recording is complete. If you want to view it later, just call it up from the Finder.

> To burn a video DVD of a Video-over-IP conference, record it with Snapz and then import the QuickTime file into iDVD.

## CHAPTER THREE

Skype and Skyping

Hacks 28–40

If you don't already use Skype, you really don't know what you're missing. Skype is the predominant desktop VoIP application: a softphone and a peer-to-peer (P2P) network that operate over the Internet to link people of all stripes around the globe. In fact, Skype has become a verb as well as a noun. You can use Skype to call people, or you can Skype people—hence this chapter's title. Oh—and you don't have to worry about finding somebody to call (that's been my problem with iChat AV), since Skype has been downloaded 150 million times and averages anywhere from 1 million to 2 million people logged in at a time.

Skype lets you make free, Internet Protocol (IP)-based phone-style calls to any other Skype user and allows you, through optional paid services called *SkypeOut* and *SkypeIn*, to place and receive calls to and from regular phones via the Public Switched Telephone Network (PSTN). Skype's sound quality is often reported to be superior to that of a traditional telephone, to boot.

Skype has several things going for it that other VoIP softphone solutions don't. It's the only P2P softphone application that runs on Windows, Mac, Pocket PC, and several flavors of Linux (Fedora, SuSE, Debian, and Mandrake, anyway). It's also the only softphone application that implements its own network and signaling protocol.

There's a reason for this proprietary characteristic of Skype's design despite all the great open VoIP standards such as Session Initiation Protocol (SIP). Since Skype uses its own P2P network and proprietary signaling protocol, it can get around the biggest problems facing the open-standard SIP protocol, which often breaks when used on phones that have to connect to the Internet through a broadband firewall router. SIP was designed without firewalls in mind, so SIP-based VoIP can prove frustrating for home users who just want to call a friend without a lot of technical hassle. By solving this problem, Skype has become the most widely used desktop VoIP application in the world. This might be why Skype's official slogan is "it just works."

Yet this is also why Skype inspires some controversy. VoIP advocates want to leverage Skype's ubiquity to advance VoIP's popularity, but to do this it needs to support the open standard for VoIP signaling: SIP. Perhaps at some point, Skype will provide for SIP compatibility or open the proprietary Skype signaling standard so that VoIP hackers can bridge the gaps between SIP-based apps and the Skype network.

Perhaps Skype's coolest feature isn't a feature at all. The Skype application programming interface (API) is a development framework that allows programmers to build *übercool* add-ons for the Skype network. I've pointed out a few of the coolest ones in this chapter.

How Skype Works

Unlike centralized voice networks like Yahoo! Chat and MSN Messenger, Skype uses a P2P network. This means that call routing is handled by a collective, serverless group of PCs running the Skype client software. As peers on the same network, each Skype node is responsible for routing calls on behalf of other nearby nodes. This gives its proprietors advantages over traditional, SIP-based VoIP telephony service providers (TSPs). For instance, Skype has less centralized infrastructure to maintain, compared to a VoIP TSP like Vonage, which has to have server capacity dedicated to every call it handles.

There are some centralized features in the global Skype network, of course. The contact search function wouldn't work so hot if it weren't able to query a global database of user information. Centralized functionality like this is clustered around Skype *supernodes*, which are actually just PCs like yours that are running Skype. Like a P2P file-sharing network, centralized search functions are facilitated using certain member PCs that are elected to have specific duties, like cataloging user data for searches and facilitating the logon process.

What It Does and Doesn't Do

Skype is largely feature-complete on all the platforms it officially supports, which makes it preferable for voice chat to something like Yahoo! Chat, which is a web-based party-line system that really works only with Windows. Skype is great for two-party direct voice calls, multiparty conference calls with up to fifty participants, and text chatting. Even cooler is the fact that Skype's user directory has advanced search functions, so you can find somebody of a certain age, gender, name, or country.

But Skype doesn't have the social networking depth of the Yahoo! system, so if you're looking for a voice chat "room" where people can freely come and go from the conversation, Skype is inferior to Yahoo! Chat. And unlike the Yahoo! and iChat instant messaging tools, there's no built-in support for videoconferencing. For that, you'll need to download and install one of the video add-ons [Hack #39].

What About Security?

As an ethically concerned VoIP hacker, you're probably wondering if this P2P network is secure. The answer is that Skype is and isn't secure. The fact that Skype encrypts all call signaling and media transmission does point to security, but the fact that your calls are routed using anonymous PCs that are members of the Skype P2P network points to lack of security. As computers become faster and faster, it might someday be trivial to crack Skype's encryption, and when that day comes, the P2P network itself will be a security flaw. For the time being, Skype is fortunately quite secure, so dig in. One hundred fifty million downloaders can't be wrong. If you need more information than what is provided in this chapter, take a look at *Skype Hacks* (O'Reilly).

Get Skype and Make Some New Friends

#28

Looking for like-minded buddies on the Net? Look no further than Skype's built-in contact search function.

You can obtain the Skype software at *http://www.skype.com/*. Though it's available for the Pocket PC [Hack #34], it is best to have your first Skype experience using a desktop operating system: Windows, Mac OS X, or Linux. The Pocket PC version is really nifty, but not entirely practical. Nor is it very customizable. So download one of the desktop OS versions from Skype's web site and install it.

Windows users will need Windows 2000, XP, or newer. Linux users will need SuSE, Fedora Core, Debian, or Mandriva (consult Skype's web site to find out precisely which kernel versions are supported). Mac OS X users will need version 10.3 (Panther) or newer. On all platforms, 256 MB of RAM is a reasonable minimum, though you might be able to get by with less.

Set Up Skype

Once you've downloaded Skype, setting it up is as simple as running its installer (Windows) or dragging its application icon to the Applications folder (Mac). On all platforms, the Skype installer is practically foolproof.

(You'll find that ease of use is Skype's middle name; for instance, you can call a person simply by double-clicking his name.) The first time you run Skype, you can set up a new Skype username, or reuse an existing one to log in. Creating a Skype account is free, though some Skype features, such as voicemail, require a paid subscription.

Find Someone to Talk To

The quickest way to find someone to talk to is to ask a real-life friend for his Skype username. Of course, if you don't have a real-life pal to Skype, you can search the Skype user directory for a lonely soul willing to talk to someone. Seriously, Skype has a mode called Skype Me that users can enable, which says they are willing to talk to anyone who comes calling. To find people willing to talk, click the Add button in Skype's main window; then click the magnifying glass icon to display the search window. Here, you'll see a checkbox labeled "Search for people in 'Skype Me' mode." Enabling this option before clicking Search will hopefully display users looking for buddies to call them. Double-click a search result to call that person.

One reason people have embraced Skype globally is because of its integrated social network. People are able to find each other through Skype's built-in contact search (available through Contacts/Search for People). If you're looking for somebody with whom you can speak German, for example, it usually takes only a single visit to this window to find users in Germany with the status of Skype Me.

Of course, if your objective is to *learn* German (or Mandarin, or Japanese... you get the idea), you can search the Skype online forums at *http://forum. skype.com/* for "learn German" or "learn Japanese." You'll find bilingual users who would love for you to Skype them, with whom you can try out your linguistic chops. A casual perusal of the Skype Me forum will put you in touch with thousands of other folks who are endpoints on the social network—endpoints with various interests:

- Pets (*http://forum.skype.com/viewtopic.php?t=29213*)
- Chinese genealogy (*http://forum.skype.com/viewtopic.php?t=29029*)
- CB radio (*http://forum.skype.com/viewtopic.php?t=28119*)

If you really want somebody to Skype you, post a message in one of these forums, and before too long, you'll have a buddy list that's a mile long (or a mile longer). Maybe you'll even learn some Swahili.

Still Don't Know Whom to Call? Meet Kerli

Here's something you'll find useful on both the desktop and mobile editions of Skype: the official (though minimally documented) Skype echo test service. This automated Skype service is a user named echo123 that records a 10-second sample of your voice and then plays it back for you to listen to. This will give you a rough idea of how well your Skype setup is working, and *whether* it is working.

> According to blog lore, the voice that is heard in the sound-test announcement is purported to belong to Kerli, a young Estonian woman.

There are also Chinese and Japanese sound-test users that you can call, each with their own pleasant-sounding announcements. To sound-check in Chinese, try Skyping *echo-chinese*. If you'd like to sound-check in Japanese, try *soundtestjapanese*.

These sound-test users offer another neat trick. If you send the text message `callme`, they'll call you to initiate the sound check. This will help you verify that your Skype is fully working—that you can place and receive calls, hear and be heard.

A Solution for Those Inevitable Antisocial Moments

Sometimes you just don't want to be bothered. That's why Skype tells you who is calling so that you can opt to ignore their calls. Then again, you might accept a call, thinking it's going to be a short one, only to hear the caller blather on about something about which they mistakenly believe you care. In times like these, you need a way out of the call. Luckily, there's Gotta Go, a Yahoo! Widget that plays a sound (like a fire alarm, for example) that gives you just the excuse you need to end the call. The calling party hears the sound, giving legitimacy to your claim that you've "gotta go!"

Check it out at *http://www.widgetgallery.com/view.php?widget=27970*.

Skype Your Outlook Contacts
HACK #29
Place Skype calls from Outlook, and even log your phone calls.

If you spend much time using Microsoft Office, Outlook is likely your email program of choice. And if you're an Outlook user, you can leverage Skype directly within Outlook, allowing you to use your Outlook address book to contact Skype buddies. All of this Office-integrated goodness comes by way of the Skype Toolbar for Outlook add-on, a program that combines Office and Skype APIs to turn your emailer into the world's coolest softphone.

The hardest part of the setup process is the download, which you can grab from *http://share.skype.com/directory/skype_toolbar_for_outlook_(beta)/*. I think it's safe to assume that once this software is no longer in beta, you can remove the *_(beta)* from that URL. Once you've downloaded the software and stepped through the installer, you'll be eager to press buttons and turn knobs.

Your New Outlook Toolbar

The first thing you'll probably notice when you launch Outlook is that it has a new toolbar that looks like the one shown in Figure 3-1. The first button on the bar launches Skype, and the second button provides a drop-down menu that lets you change your Skype status (very handy) and export your Outlook contacts to Skype's buddy list.

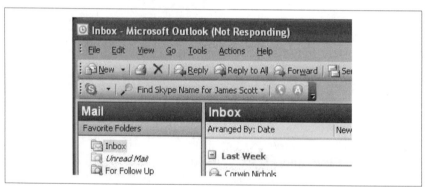

Figure 3-1. The Skype Toolbar for Outlook

This export process, pictured in Figure 3-2, is a little shaky, but it does its job of matching users from the global Skype directory to your existing Outlook contacts with passable accuracy, though at a snail's pace (it took my batch of only 48 contacts about 10 minutes).

The next toolbar item changes depending on the contact of the message or the contact selected. Say you're navigating your Inbox and you open a message from somebody whose phone number is in your address book. This button changes to show the phone number of the person who sent the mail and enables you to Skype-call that contact (via SkypeOut) with one click. If the contact is in your Skype buddy list, this button will call the contact via his Skype name. If you have neither a phone number in your address book

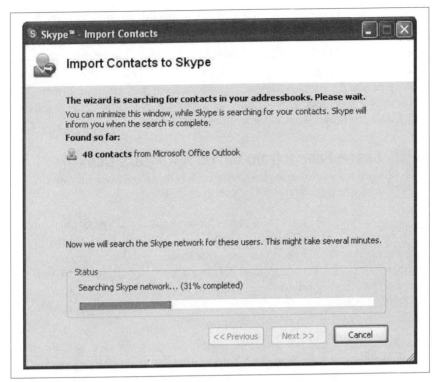

Figure 3-2. Exporting your contacts to Skype

nor a buddy list entry for this contact, the button will attempt to look up the contact's Skype name via the Skype directory. So you're covered in any event, unless your contact doesn't have a telephone or a copy of Skype installed. The last button on the toolbar, an *A*, lets you launch a text chat with the current contact, too.

Journal Your Skype Calls

What could be more useful than tracking when and whom you Skyped? This capability is actually built into Skype, via its call list feature. But using Outlook's Journal feature to track the whens and whos of Skype is far more useful. The Journal puts Office events (and now Skype events) in a chronological, searchable view that resembles a timeline and provides a sort of audit trail for changes made to files, messages, and contacts within Microsoft Office. Long after a particular message, contact, or Skype call has been deleted, the journal still retains a record of it—when it happened, who called (or who you called), and how long. But the Journal option for Skype isn't enabled unless you tell it you want it—an option you'll find when you click the Configuration option located in the drop-down menu on the Skype Toolbar for Outlook.

If you'd like a slightly different approach to saving a log of calls made and received with Skype, try Avantlook (*http://share.skype.com/directory/ avantlook/view/*), another Skype toolbar add-on for Outlook. Instead of using the Office Journal, it actually stores Skype events as items in its own searchable, sortable message folder.

Skype People from the OS X Address Book

HACK #30

Using AppleScript and the Mac OS X Address Book, you can Skype any phone number in your address book with two mouse clicks.

Though the Skype API provides some power tools for Skype hackers on the Mac, you can do quite a bit with Skype using plain-old AppleScript. This is great, since not all hackers prefer the same tools (or, if you're like me, you've stubbornly refused to understand C). In "Skype Your Outlook Contacts" [Hack #29], you saw a few different ways to tie Windows-based Outlook to Skype. You can do the same thing for the Mac OS X Address Book with a little bit of assistance from trusty old AppleScript.

Thanks to the simple but awesome script shown in the next section, it's a snap to call any of your Mac OS X Address Book contacts using Skype.

The Code

For starters, you'll want to key the following script into the AppleScript Editor. You can find the AppleScript Editor in your Mac's Applications folder:

```
using terms from application "Address Book"
    on action property
        return "phone"
    end action property

    on action title for p with e
        return "Call with Skype"
    end action title

    on should enable action for p with e
        return true
    end should enable action

    on perform action for p with e

        set x to value of e as string
        if character 1 of x is not "+" then
            set x to "+1" & x
        end if
```

```
    set SKYPEurl to "callto://" & x
    tell application "Skype"
        get URL SKYPEurl
        activate
    end tell
    return true
end perform action
end using terms from
```

Running the Code

In the AppleScript Editor, save this code as an AppleScript in your user's *Library/Address Book Add-ins* folder. Then, fire up the OS X Address Book and find a contact with a phone number, as shown in Figure 3-3.

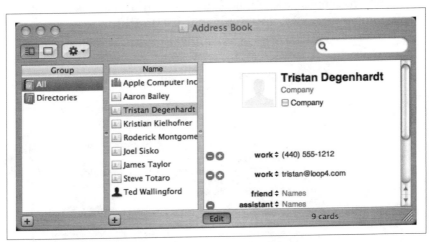

Figure 3-3. The Mac OS X Address Book

Now, right-click (or Ctrl-click) the contact's phone number. The contextual menu should contain an option labeled Call with Skype if you've done everything correctly up until this point. Clicking that option will cause Skype to attempt a call to the phone number in the contact record. Note that you'll need to have SkypeOut credit to call a PSTN phone number using Skype.

> Remember those Skype pay services I mentioned at the beginning of this chapter? SkypeOut is one of them. It lets you call old-fashioned phone numbers from your Skype client for just a few cents a minute. You can buy minutes ahead of time from *http://www.skype.com/*.

Even though the Address Book's phone fields are referred to as numeric in nature throughout the Address Book AppleScript documentation, you can enter alphanumeric values for any phone field. So, if you want to Skype somebody from your address book using their Skype name rather than their PSTN phone number, just enter their Skype name into a phone number field in the Address Book. That will work just fine.

HACK #31 Enable Site Visitors to Skype You

Want to make it easier for folks to Skype you? Put a Skype Me link on your web site or blog, and your visitors will chime in.

Chances are if you have a web site, you have more than enough HTML skills to pull off this simple hack. If visitors to your web site have Skype installed on their computers, they already have the Callto URI handler set up—in other words, they can click a link on your page to call you using Skype. Here's an example of such a link, at the upper-left corner of the page in Figure 3-4.

Figure 3-4. A Skype Me link embedded in a web page

The nifty Skype Me graphic is provided by the good folks at Skype, so all you need to do to create this link is plug in a bit of HTML with your Skype username, like this:

```
<a href="callto://Your_Skype_Name">
<img src="http://goodies.skype.com/graphics/skypeme_btn_small_blue.gif"
border="0"></a>
```

You'll need to replace *Your_Skype_Name* with your own Skype name—hopefully that was obvious to you. The image I've used in Figure 3-4 is blue (believe it or not). If blue isn't your color, you can visit *http://www.skype.com/community/* for a host of other colors, sizes, and styles of ready-made Skype Me graphics. Once you stick one of these buttons on your home page, all Skype users need do is click it, and seconds later, you'll receive a Skype call.

> If the user clicking your Skype link doesn't have Skype installed and is on a Windows machine, the Callto handler will try using NetMeeting instead! For this reason, you should add a Skype download link for them and suggest visitors install it before they try to Skype you using your link.

HACK #32   Speak Jyve

Skype is not just a communications tool; it's a social-networking platform, too.

Skype inherits many of the cool features of instant messaging, including *presence*, that useful list of status names that your buddies can use to see whether you're available to chat. This includes status indicators that tell if you are Away, Busy, or Available—and, of course, Skype Me, which allows you to connect to people around the world interested in chatting with strangers. But wouldn't it be cool if you could put a status indicator on your web site so that folks would know whether it's a good time to Skype you?

Thanks to Jyve, you can do precisely that. Jyve is a social-networking service for Skype that adds some pretty useful features to an already great application. (If your social life is in need of new features, though, you might need the input of something other than Jyve.) Aside from being able to publish your status on your home page, visitors to your page will be able to request a phone call (a Skype or old-fashioned phone call) from you via a text message that is sent to your Skype client. That way, even if you aren't there to chat, the message will be waiting on your screen when you return.

Jyve even provides the equivalent of a V-Card contact record for your web site, which Jyve calls a *Q-Card*. It's a web page hosted by Jyve that contains your Skype information and a form that visitors can use to text-message you (even if they don't have Skype). Using a bit of JavaScript pasted onto your home page, you can integrate your Q-Card into your web site.

> A *V-Card* is a small file that contains contact information for a certain person, like a "virtual business card." V-Cards are used by email programs, for the most part.

Get Signed Up

Does all this web-enabled Skype goodness sound like fun? I'm sure it does, but first things first. To use Jyve's service, you've got to create an account on the Jyve server. To do so, surf over to *http://www.jyve.com/* and find the "Get your Skype Card Here" link on the main page. Here, you'll have to fill out a pretty standard membership form. On the following page, you'll be able to create and customize your Q-Card. You can preview your Q-Card, too. Though it's optional, you should upload some kind of portrait; the Q-Card looks a bit bare without a picture.

Make Jyve a Buddy

You might be wondering how Jyve is able to keep tabs on your Skype status, since only Skype buddies can see your Skype presence information. You're completely right—only Skype buddies can see if you're "Available" or "Hacking your Volkswagen." So, to share this info with people who view your Jyve Q-Card, you'll need to make Jyve a buddy in your buddy list. To do this, click the Add button and type **jyve01** for the buddy's Skype name (or type what Jyve instructs you to type on your account page).

Add Jyve's HTML to Your Web Page

On your account page, click the Get HTML button, and you should get a text box with some HTML that looks like this:

```
<a href="javascript:void(0)" onClick='window.open("http://jyvesolutions.
jyve.com/Qcard/your_name.htm",
"QCard","menubar=no,scrollbars=no,height=500,width=800")'>
<img src="http://jyvepresence1.com/qzoxy/your_name.png"
border="0"></a>
```

This bit of HTML, when placed in the source of your own web page, will put your portrait on the web page with a link to your Q-Card, which will open in a new window when clicked. Figure 3-5 shows you an example of a Q-Card embedded in a web page using the preceding HTML.

Start Jyving

The Q-Card in Figure 3-5 is being used to send a text message. This is ideal for enabling people to get in touch with you, even if they don't have Skype, because they can enter their call-back number, and it will be sent to you in the text message. This could be a great customer-service tool for your company web site, or just a nice novelty item for your personal blog. Figure 3-6 shows what an incoming Jyve text message looks like.

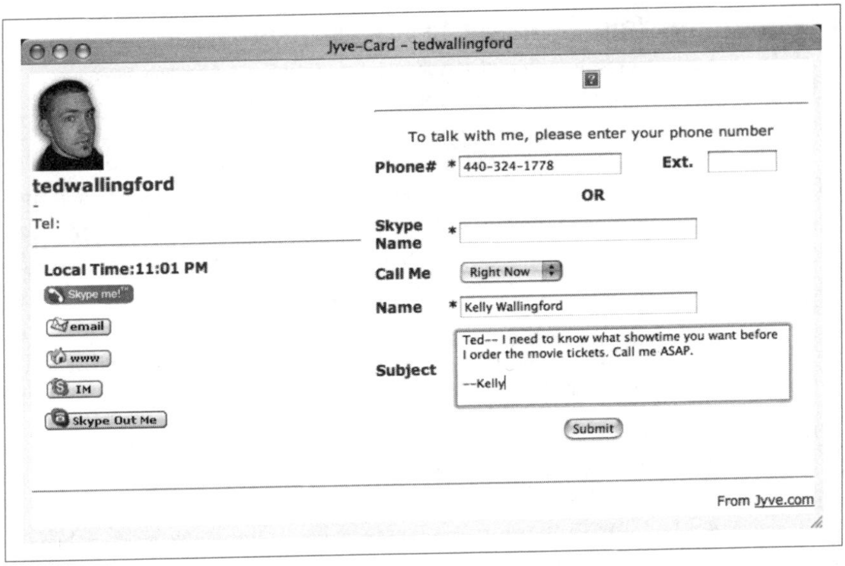

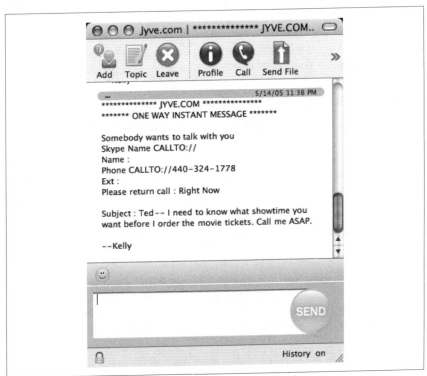

Figure 3-5. A Jyve Q-Card

Figure 3-6. A Jyve text message sent from the Q-Card

HACK #33 Teach Your Browser to Speak Jyve

With a little extra help from the Jyve web browser plug-in, you can extend Skype's presence features to your web browser.

If you've signed up for a Jyve account, you already know how cool the Jyve Q-Card is. But with the Jyve web browser plug-in, you can take Jyve's intimate use of Skype's presence features to the next level. This might mean being able to receive a Skype instant message whenever somebody visits a certain page on your web site, or even letting people send you a voicemail from your web site. Unfortunately, to enable these features, your web site visitors must install the Jyve browser plug-in, which works only with Windows. You can get the plug-in at *http://www.jyve.com/*.

Once installed, the Jyve plug-in enables a number of new protocol prefixes that allow your web browser to trigger different kinds of Skype functionality. These prefixes include `IMTO:`, `SVMTO:`, and `CREATECONFCALL:`. Using these prefixes and a bit of HTML, you can create links on your personal web page that leverage some of Skype's coolest features, like instant messaging and voicemail.

 Like SkypeOut and SkypeIn, Skype Voicemail is a paid service. To use voicemail features, you'll need to have paid for Skype Voicemail service.

Add Skype Instant Messaging to Your Web Site

To create a link on your site that allows surfers to send you an instant message, use the `IMTO:` prefix and the following code:

```
<script src="http://plugin.jyve.com/js/plugin.js"></script>
<a href="IMTO://voiphacks~Hello World"><img src='http://jyvetools.jyve.com/
IMTO.gif' border='0'
onclick='setENDown();'></a>
```

Track Visits to Your Site by Jyve Users

You can create a trigger on a web page that sends text messages to you when the web page is visited, without allowing the surfer to compose or add to them. The visitor won't see an instant message window, either. The idea here is to send you a message when somebody hits a certain page on your site. Using the following JavaScript in one of your page body's onLoad events, you can use a "pop-under" window to send you a Skype instant message automatically whenever a visitor with the Jyve plug-in visits:

```
<script language="JavaScript">
<!--
function popunder() {
```

```
pu = window.open("SENDIMTO://voiphacks~I visited your site!", "default");
pu.blur();
window.focus();
pu.close();
}
//-->
</script>
<body onLoad="javascript:popunder()">
```

Simplify Communication for Visitors to Your Site

The best thing you can do if you want to keep in touch with somebody is get into their buddy list and vice versa. If you'd like to have a link that lets the surfer add you as a contact to his Skype contacts list, try a link like the following, which pops up a dialog confirming the addition of you as a new contact in her list:

```
<script src="http://plugin.jyve.com/js/plugin.js"></script>
<a href="ADDCONTACT://voiphacks"><img src='http://jyvetools.jyve.com/
addcontact.gif' border='0' onclick='setENDown();'></a>
```

Finally, to make a link that allows surfers to send you a Skype voicemail (sorry, the Skype Answering Machine, a free voicemail alternative to Skype's official Voicemail service, won't work with the Jyve SVMTO: prefix), try a link like this one:

```
<script src="http://plugin.jyve.com/js/plugin.js"></script>
<a href="SVMTO://jyvetest1"><img
src='http://jyvetools.jyve.com/sendvoicemail.gif' border='0'
onclick='setENDown();'></a>
```

If you don't like the premade graphics that Jyve provides (in the <img> tags), you can use your own or substitute text instead.

Trigger Conference Calls from the Web

Skype also includes a function that lets you create a link that starts a conference call with the Skype users of your choosing. This might be useful for somebody who hosts a regular conference call with the same people every week—like an editor and her authors. By keeping a central link on a web page that all the attendees can access, the weekly conference can still go on, even if the regular host isn't available to set it up. A fill-in member need only click the link to launch the conference call. Try a link like the following:

```
<script src="http://plugin.jyve.com/js/plugin.js"></script>
<a href="CREATECONFCALL://jyvetest1~garfield~odie~nermil"><img
src='http://jyvetools.jyve.com/sendvoicemail.gif' border='0'
onclick='setENDown();'></a>
```

Just put a tilde (~) between each contact in the link. When clicked, the Jyve plug-in will use the Skype client of the clicker to host the conference call.

Carry Skype in Your Pocket

#34 Turn your Pocket PC PDA into a wireless VoIP phone.

You can use your WiFi-enabled Pocket PC as your personal Skype portable phone. Well, not any Pocket PC; only a Pocket PC running Windows Mobile 2003 will work. And it'll need a 400MHz processor, too. Pretty hefty specs, I know. But to do cool things, you need cool hardware, right? (The sizzling HP iPAQ hx4700 Pocket PC has an XScale processor that cruises along at a cool 624 MHz, incidentally.)

Grab PocketSkype from *http://www.skype.com/products/skype/pocketpc/* and install it on your Pocket PC device using ActiveSync. Once you've installed it, you should be able to Skype people using the Pocket PC in the cradle, sharing your PC's network connection. The PocketSkype user interface, shown in Figure 3-7, is similar to the desktop versions.

Figure 3-7. Skype for Windows Mobile 2003

PocketSkype works equally well uncradled, provided you have an 802.11b or better wireless connection. This is plenty of bandwidth for Skype calling, but keep in mind a couple of things:

- Using wireless will drain your Pocket PC's battery.
- Wireless LAN will add some delay to your voice signal transmission, often resulting in noticeable lag during the conversation. (To test the sound of your Skype calls, read "Get Skype and Make Some New Friends" [Hack #28].)

- The range of most wireless LANs is a few hundred feet (less than that indoors), so your PocketSkype will only be as mobile as your LAN allows.

Pocket PCs aren't known for their stunning wireless reception, so if you plan to rely on your PocketSkype setup, invest in an extended-range base station. Try to use a channel that performs very well (try pinging over your wireless LAN and watch for *jitter*—big variances in the ping time; this is bad), and locate the base as close to the center of its desired range as possible. If you can't cover the entire area, connect a WLAN extender like Apple's AirPort Express in the area the base station doesn't cover. Or, use a couple of base stations.

Don't Forget Wireless Security

To make sure your neighbor can't inadvertently access your wireless LAN from his patio, and possibly make his own Skype calls on your dime, you'd better secure your wireless access point. This is good practice even if you don't use Skype.

There are three ways to do this. Either restrict the MAC hardware addresses allowed to connect to your base station, or use a Wireless Encryption Protocol (WEP) shared key to keep your neighbor's hands off your wireless LAN. The third way is to use both techniques. This is the most secure method of all.

To find out the MAC address on your Pocket PC, power it up and tap Start. Then tap Settings, and then the System tab on the bottom of the screen. Next, tap the WLAN icon (it might say "iPAQ WLAN" depending on your brand of Pocket PC). Then tap the Status button. The MAC address will be listed near the bottom of the screen. Add it to your wireless base station's access list.

Degunk International SkypeOut Calls

With SkypeOut, you can easily make international long-distance calls, but keeping track of international country codes isn't so easy. Neither is keeping track of calling rates, which vary from one country to the next.

The SkypeOut service is one of several paid services (SkypeIn, Voicemail, etc.) that Skype provides for its users. The service lets you place calls to regular phone numbers on the PSTN—that's the telephony network still used by everybody who isn't as cool as you are.

SkypeOut minutes are purchased in advance, and, at least for international calls, their rate is quite competitive. You can SkypeOut to any existing phone number—your buddy's cell phone in Uruguay or dear old mom in Kalamazoo. The trick isn't merely calling folks; it's remembering how to dial their numbers. If you already do a lot of international calling, you're probably familiar with the system of country codes that exists to route calls around the globe. Country codes are dialed before area codes as a part of the receiving party's phone number. But there are almost as many country codes as there are countries, so you might appreciate this shortcut to help you remember how to dial international numbers on Skype.

At *http://www.skype.com/products/skypeout/rates/*, Skype maintains a nifty script that allows you to enter the local phone number (including area code) and select the country. Then, the script shows you exactly how to dial it using Skype, as in Figure 3-8. Just follow the steps on the page: enter the phone number and pick a country. Besides learning how to dial the number, you'll also get a handy SkypeOut rate quote for calls to this number (in euros).

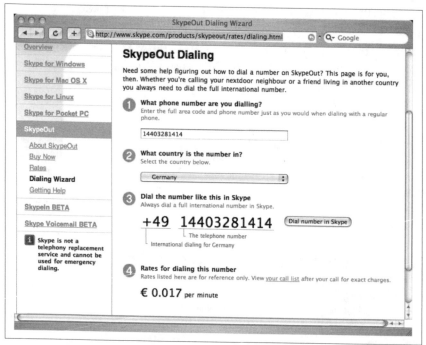

Figure 3-8. The SkypeOut Dialing Wizard

To convert from euros to the currency of your choice, you've got a couple of options. You can go the web-based route, using a site like *http://www.xe. com/ucc/*, the Universal Currency Converter. Or, if you're a Mac OS X 10.4 (Tiger) user, you can launch the handy Unit Converter widget in the

Dashboard. For more information about the Dashboard and other cool Mac goodies, pick up a copy of *Mac OS X Tiger: The Missing Manual* (O'Reilly). Whether it's currency conversion for SkypeOut or just checking the weather forecast, widgets are really nifty. Windows and Mac users alike should take a look at Yahoo!'s really cool widget framework (formerly known as Konfabulator) at *http://widgets.yahoo.com/*.

From Podcasting to Skypecasting

With some simple recording tools, you can easily integrate Skype—and other VoIP calling software—into your podcasts.

Audio blogging, or *podcasting*, as it's called, is a form of amateur radio broadcasting that uses the Web as a means of distribution, just as blogging is a personal form of journalism that uses the Web for distribution. Podcasters record their broadcasts in MP3 audio files, and their audience downloads them for listening on a PC, iPod, or other portable MP3 player.

> The art and science of podcasting deserves its own book. Fortunately, a great one is available. For a more detailed explanation of podcasting, check out *Podcasting Hacks* (O'Reilly).

If You Build It, They Will Come

Creating cool content for podcasts isn't always easy. Without a broadcast license, it's not a simple thing to use (legally, anyway) copyrighted music in podcasts, since their distribution is heavily protected by the long legal arm of the Recording Industry Association of America (RIAA), the same folks who dumped millions into suing harmless teenagers who download Coldplay singles that sell for 99 cents apiece through iTunes. But I digress.

The point is that podcasts can get old really quickly unless you have a constant flow of good audio content, and this doesn't come easily. Fortunately, the talk format, which is stridently avoided by most of those downloading teenagers I mentioned earlier, is a content miracle. It won't get you in trouble with the RIAA, either.

But you don't need to have Rush Limbaugh's golden microphone or Howard Stern's subtle wit to master the talk format in your podcasts. You just need to be able to record your Skype (or MSN Messenger, or AIM) VoIP conversations, call them "interviews," and podcast them! You can also use your computer to record your traditional phone calls for podcast purposes, using tools like Phlink [Hack #15].

Mac Podcasting Tools

Using a tool like WireTap [Hack #23], grabbing the audio from a Skype conversation is a snap. Then it's just a matter of editing it down (the talk format isn't forgiving of coughing fits or belches) and converting it to MP3 format. Using Cacophony or GarageBand, you can insert the recorded interview into the middle of a podcast, or give it some bumper music just like the pro talkers.

Another cool app for recording audio from VoIP conversations on Mac OS X is Soundflower (*http://www.cycling74.com/products/soundflower.html*). This app actually lets you treat the output of any application as a sound input device in other applications, so you can use it to record Skype conversations and to process them later on so that they sound differently. With GarageBand, for instance, you can add reverb to a live sound input before adding it to the podcast. Or you can run it through EQ, compress it, or do a host of other cool stuff to it to whip it into shape for your own radio show.

Great for sound effects, GarageBand, in concert with Soundflower, could be the makings of the ultimate radio drama podcast. You can apply reverb, echo, and pitch shifting to sound inputs in GarageBand, taking your dramatic podcast to the Grand Canyon, or to a village of squeaky-talking munchkins as in *The Wizard of Oz*.

Windows Podcasting Tools

A great place to start on your quest to make the ultimate Skypecast is Total Recorder Standard Edition, a Windows shareware tool that you can grab from *http://www.highcriteria.com/*. Like its Mac counterpart, WireTap, Total Recorder lets you intercept audio from one or more channels, like the sounds from a Skype conversation, and save it to a file that you can integrate into your Skypecast.

As you can see in Figure 3-9, the creators of Total Recorder were thoughtful enough to create the "Record also input stream" option, which automatically mixes the microphone input channel with the sound output you're recording, simplifying the task of recording your Skypecast (that is, unless you don't want your voice to be heard during the interview). Once you've saved a WAV file with your interview, you can edit it into the rest of your podcast using Windows Sound Recorder or your favorite sound editor.

Then, if your editor doesn't support saving the whole thing as an MP3, you can run the finished WAV version through SoX (see Chapter 2 for a refresher on SoX) to make it ready for publishing to the Web:

```
$ sox my_skypecast.wav  my_skypecast.mp3
```

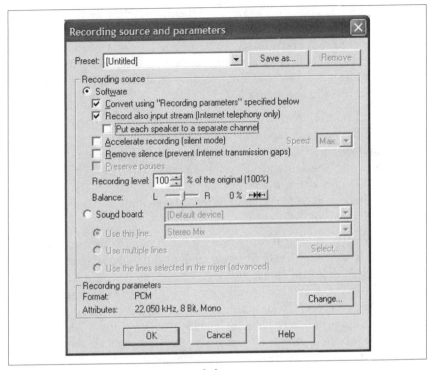

Figure 3-9. Total Recorder's parameters dialog

Chances are that your Linux box already has SoX installed. If not, or if you don't have a Linux box, you can grab SoX for Windows at *http://prdownloads.sourceforge.net/sox/sox12177.zip?download*.

Three's a Crowd

One of Skype's knock-dead features is its multiparty conference-call support. Up to fifty participants can talk together, just like a traditional conference call. By recording a conference call, you can create a podcast-friendly "panel discussion." Even better, if you use Alex Rosenbaum's Skype Answering Machine software to greet callers while you're away [Hack #37], you can even *use* Skype to do the broadcasting, via the greeting the Skype Answering Machine plays for callers as they Skype you. Check out the next hack for the details.

Experiment Your Way to a Perfect Skypecast

A lot of variables are involved in recording a podcast. Many of the fundamental techniques of voice recording carry over from the radio broadcasting industry. The human voice is very dynamic. It's very quiet one moment, and very loud or boomy the next. Yet, the human ear prefers a much more controlled aural dynamic—that is, we'd rather listen to the human voice at a consistent volume level, especially when there's background noise. That's why professional radio broadcasts are *compressed*, or processed so that the quiet moments are louder and the loud moments are quieter. This way, the sound is at a much more consistent volume level (traditional phone calls work in much the same manner).

Microphones matter. A microphone with a *windscreen* (that afro-shaped sponge you see on news reporters' microphones) will reduce the unprofessional-sounding harshness caused when pronouncing the letters *f* and *p*. A microphone with a wide frequency response range will pick up more of your voice than a laptop's built-in mic or a cheap $9 USB mic. Of course, a good mic on *your* desk won't help your interviewee sound any better. They'll need a good mic, too.

Using different voice chat tools (iChat, Skype, Yahoo! Chat, etc.) will certainly deliver different levels of compression and equalization, and you might discover that you prefer one over the other. Using different post-processing tools (GarageBand, AcidPro, etc.) will afford you greater flexibility in making your podcast sound the way you want it to. But don't expect any of these applications to have the magical "podcast preset." You'll need to experiment until it sounds like what you think a broadcast should sound like. If you're talking a lot, listen to talk radio for inspiration (not only with respect to production, but also the subject matter).

See Also

- *Podcasting Hacks* (O'Reilly)

HACK #37 Answer Your Skype Calls, Even When You're Not Around

Thanks to the Skype API, Windows and Mac developers can create useful tools to extend the functionality of the Skype platform. One such tool is the Skype Answering Machine for Windows.

Skype makes you ultimately accessible. In fact, if you log into your Skype account using Skype clients on three different computers (on three different continents) simultaneously, any Skype calls to you will alert you on all three computers at once. You can answer such calls on any of the three, to boot!

But sometimes you might not want to be bothered. Say that you're out on your million-dollar pleasure yacht in some exotic port being fed plump red grapes while lounging behind your Ray-Bans. Now would not be a good time to receive a Skype alert (unless it is from the ship's master chef, informing you that your filet mignon is ready).

Times like this call for the Skype Answering Machine (SAM), at least if you're a Windows user (though the Skype API exists for Mac, SAM is only for Windows). This software add-on for Skype is a fully featured answering machine that can record your callers' messages, and even greet unknown callers differently from callers that are already in your buddy list. Grab it from *http://www.freewebs.com/skypeansweringmachine*.

> Any time you add a new Skype add-on, like SAM, Skype will prompt you to grant permission for the new add-on to access Skype. This is a security precaution that's built into Skype, so don't be alarmed. If you've used Windows for any length of time, you're probably quite accustomed to these security warnings.

It's a tiny download. Close Skype before you run the installer. Once it's installed, you'll see a small, green SAM icon in your system tray. Double-click it to launch the user interface, and the first thing you'll see is SAM's call-log dialog. This is where each of your calls will be logged and you can listen to the messages that folks leave you. Click the Options tab, as shown in Figure 3-10, to see where the fun settings are.

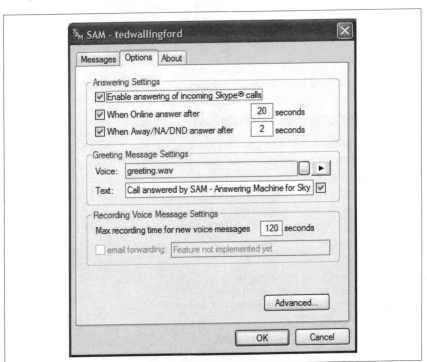

Figure 3-10. Skype Answering Machine's Options tab

You can tweak the answering delay time and select your own recording to use as the outgoing greeting. SAM supports only WAV files, but you can record those easily enough using the Windows Sound Recorder. But since SAM also answers calls with a text message, you can specify the text to use as your "auto-away" message. You can also set the maximum length of incoming recordings. The Advanced button will allow you to establish a second configuration set for calls originating from folks that aren't in your buddy list, as shown in Figure 3-11.

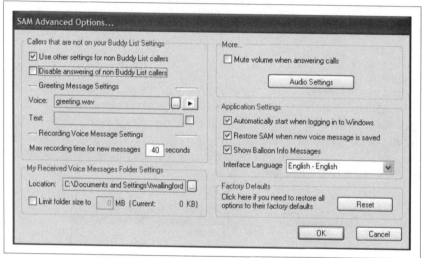

Figure 3-11. Skype Answering Machine's Advanced Options dialog

Let Windows Automatically Rotate Your Greetings

If you want to use a different greeting depending on the time of day, similar to the daily rotation used in "Greet Callers Differently Each Day" **[Hack #17]** you can use a batch file combined with Windows Scheduled Tasks to rotate the greetings at predetermined intervals. In this case, let's do an a.m./p.m. rotation, meaning we'll need just two greetings. You can set up a single Windows batch file to swap two previously recorded greeting files stored in a *\greetings* directory:

```
cd \greetings
copy /Y greeting.wav greeting_tmp.wav
copy /Y greeting_bak.wav greeting.wav
copy /Y greeting_tmp.wav greeting_bak.wav
del greeting_tmp.wav
```

Now, copy your p.m. greeting to *greeting.wav*, and we'll schedule this to run beginning at noon using Scheduled Tasks, available from Start → All

Programs → Accessories → System Tools → Schedule Tasks. Set the task to start at 12:00 p.m., and enter the full path to the batch file you created earlier. Click Advanced, and you'll see a dialog like the one in Figure 3-12.

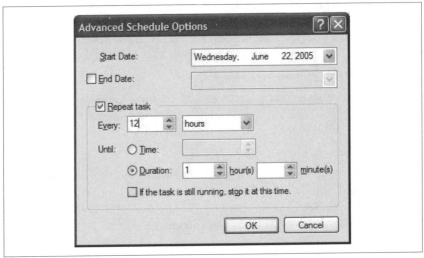

Figure 3-12. Windows Scheduled Task's Advance Schedule dialog

Check the Repeat Task checkbox, and set this task to run every 12 hours. Then click OK, and click OK again on the preceding Scheduled Tasks window. Now, every day at noon and midnight, beginning at noon today if it's morning, your Skype Answering Machine greeting files will be swapped.

Skype's Voicemail Service

If you're using Skype on a platform that doesn't support the Skype API, or if you would prefer a voicemail solution that allows you to retrieve your voicemail messages from anywhere (not just from your Skype PC), you should consider Skype's subscription-based Voicemail service. You can sign up for the service at *https://secure.skype.com/store/member/login.html*.

HACK
#38 Use Custom Rings and Sounds with Skype

Skype allows you to use your own sounds to alert yourself to incoming calls and events—with a few gotchas.

One of the most popular downloads (and purchases) for cell phones are customized ringtones. This feature is available for Skype too, with none of the proprietary nonsense pushed on you by your cell-phone carrier; you can use just about any WAV file as a Skype ringtone. (On a Mac, an AIFF file will suffice, too!) But not just any WAV file will make Skype happy: stereo

WAVs won't work, as the WAV needs to be monaural (one channel). Thankfully, you can use a common recording tool to convert your stereo WAV file so that it can work with Skype.

If you've got SOund eXchange (SoX) installed (see "From Podcasting to Skypecasting" [Hack #36] to find out where to get the Windows version) on your Windows, Mac, or Linux machine, you're ready to convert files to mono:

```
C:\> sox stereo_file.wav -c 1 mono_file.wav
```

SoX's -c option knocks that stereo file down to one channel, perfect for use with Skype. If you'd like to convert an MP3 file into a mono WAV file, just specify the WAV file extension for the destination file's name:

```
C:\>sox stereo_file.mp3 -c 1 mono_file.wav
```

Drop the file created by SoX into a directory where you can access it from Skype, and pull up Skype's Options dialog. To do so on Windows, select Options from the Tools menu in Skype (Figure 3-13). To do so on a Mac, select Preferences from the Skype application menu (Figure 3-14).

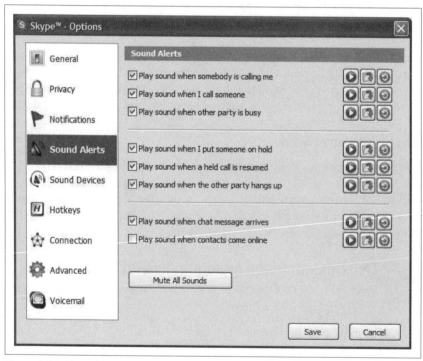

Figure 3-13. Skype for Windows Sound Alerts options dialog

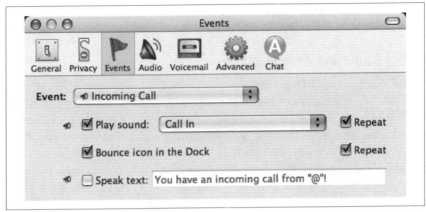

Figure 3-14. Skype for Mac OS X Events options dialog

When the preferences window appears, click Sound Alerts (Windows) or Events (Mac). Here, you can browse to the sound file you'd like to assign as your ringtone. The Mac version even allows you to have Skype use OS X's speech synthesizer to announce the name of the incoming caller, if you like.

It's not just the incoming call event that you can disassociate from the default alert sound, either. Thumb through the event list to check out all the possible combinations.

Emote by Sight and Sound with Skype

HACK #39

In this modern age of voice and video, emoticons are not forgotten and have become equally modern.

In the early days of text-based chat, we developed text expressions like :) and =(to show signs of happiness and dissatisfaction, respectively. We called these *emotion indicators*, *emoticons*, or simply *emotes*. And unless you've never chatted online, never talked to an Internet user, or never seen the film *You've Got Mail* with Meg Ryan and Tom Hanks (or one of its several knockoffs), chances are good you know what emoticons are. Today, with voice chat augmenting text chat and video chat and with webcams becoming commonplace, the old-fashioned emoticon seems plain out of place. It's small, monolithic, and, well, underwhelming. Emoticons just don't convey as much emotion as a speaking voice or the human face, which is what you hear or see when you Skype folks using voice or video facilities.

You're about to use Skype add-ons to enable audio emotes, 3-D avatars, and video chat, making old-school text emotes seem downright archaic. The tools I've looked at are all built around the Skype API for Windows, though, so if you've got a Mac (for which the Skype API has only just been introduced), you'll be watching the action from the sidelines.

Adding Sound and Video Emoticons

Sound emotes and 3-D avatars are two really cool add-ons for Skype that can enhance the social aspect of the Skype experience. Sound emotes are essentially just prerecorded sounds that you trigger as a part of your normal sound transmission so that the person on the other end of the conversation can hear them. As with old-fashioned text emoticons, sound emotes can be just the thing you need to lighten up a conversation, or just to raise the silliness level a bit. (Yes, even I, the ever-stodgy VoIP aficionado, have been known to be silly once in a great while.)

To get started with sound emotes, you'll need to pick up a copy of Porto Ranelli's HotRecorder 2.0 for Windows (*http://www.hotrecorder.com/*). This ad-supported shareware application lets you select from a small batch of prerecorded sound emotes, including applause, a room full of people laughing, a baby crying—in other words, a wisecrack for every occasion. Also included in HotRecorder is a voicemail utility for Skype (I prefer Skype Answering Machine, though) and a sound recorder, so you can add your own sound emotes. To play them back during a Skype conversation, just click the one you want from the selection on the Emotisounds tab in the HotRecorder application. For some real fun, try importing sound clips from your favorite movies.

That's just sound, though. To bring the visual aspect of emoticons into the 21st century, download a copy of 3D Avatar Messenger (*http://share.skype.com/directory/skype_3d_avatar_messenger/view*) and install it on your Windows PC. It's kind of hard to describe what this application does, though Yahoo!'s IMvironments are probably the closest analogy. Figure 3-15 shows the application's interface.

3D Avatar Messenger is a Java application that uses the Skype API to send animated, three-dimensional emoticons involving a cartoon character that has several characteristics that you can manipulate—hair, shirt, and pants color, and gender (though it appears you'll be stuck with red shoes no matter what). The coolest part about 3D Avatar Messenger is how it displays your character in the same window as your conversation partner's character, allowing you to interact with him. The application is limited to two participants at a time, of course, and it's the most fun to use during a Skype voice conversation.

To install it, unpack the zip file you'll find at the URL mentioned earlier and execute its *run.bat* file,· either by double-clicking or executing at the

Figure 3-15. Skype 3D Avatar Messenger enhances VoIP calls with an animated alter ego

command prompt. This will launch the Java interpreter and allow the program to run. Once it's up and running, you'll need to find a partner who also has it installed—a husband or wife will work well (hey, it works for me). Otherwise, you'll be emoting by yourself, as in Figure 3-15.

Sometimes There's No Substitute for Video

If you want to use your webcam to enhance your Skype calls with real-time video, just like the video viewer screens on Star Trek* (only lower resolution), point your browser to two of the coolest video-on-Skype plug-ins:

- Festoon (*http://www.festooninc.com/*)
- video4IM (*http://www.video4im.com/*)

At the moment, both work only with the Windows version of Skype, but considering the recent release of the Skype API for the Mac, I suspect we'll see some video goodies for Mac Skype very soon.

To be seen by your Skype buddies, you'll need a webcam connected. Logitech, Microsoft, Sony, and Creative all sell USB webcams that are suitable

---

* Star Wars holograms look cooler than Star Trek video screens, but I'm afraid there aren't any hologram plug-ins for Skype yet.

for Skype video. You don't need a webcam to see the video from your buddy's webcam, of course.

HACK #40 Skype with Your Home Phone

Combine high-tech VoIP with a low-tech analog phone to make VoIP palatable to even your technology-phobic spouse.

If you've worked with Skype for very long, you've probably become accustomed to its (mostly) good sound quality and friendly interface. However, I always have a hard time with how it feels to be speaking to a computer. Instead of the secure feeling of an old-school phone receiver, I am uncomfortable speaking into a USB headset or, worse still, speaking into my PowerBook's built-in microphone. For once, I just wish I had a good old-fashioned analog phone to slide up next to my ear—yes, even for Skype calls.

Fortunately, this is now possible. Using the Actiontec Internet Phone Wizard, a USB device for Windows PCs, you can integrate Skype with your old-fashioned residential-style phone. The device is sort of an analog telephone adapter (ATA) that allows your PC to act as a gateway between the analog phone and the Skype network. Too cool.

Make the Connection

For this hack, you'll need a computer running Windows 2000 or newer and the Actiontec Internet Phone Wizard, pictured in Figure 3-16, which needs to connect to your PC's USB port. The two RJ11 ports on the device's back allow you to connect it to an analog phone and a plain-old telephone line. This pass-through connection allows you to place certain calls on your ordinary phone line if you prefer. That connection can also handle emergency calls (911), which is important, since Skype has no provision for 911 call routing.

Perhaps the wizard's most valuable characteristic is the way it translates Skype's features so that a traditional telephone can use them. For instance, when you receive a Skype call, the phone will ring and you can answer it; later, while you're still on that call, if you receive a second Skype call, your phone will use a call waiting signal to let you know another call is ringing in. This way, you can switch between two Skype callers as you would with call waiting on a legacy telephone line.

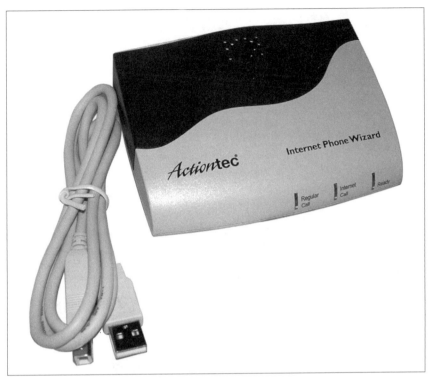

Figure 3-16. Actiontec's Internet Phone Wizard

Also supported are conference calling and speed-dial integration. This way, you still can access your Skype buddies who have alphanumeric names by dialing your telephone keypad, which has only telephone numbers. The wizard's included software lets you associate speed-dial numbers with contacts in your Skype buddy list, simplifying the act of calling them. Once you've run Actiontec's installer, your Skype buddy list will have an additional option in its contextual (right-click) menu: the Assign Speed-Dial option, shown in Figure 3-17. Click this option to define which two-digit speed-dial number to associate with each member of your buddy list. That way, when you want to Skype them using the attached phone, you need only press the speed-dial numbers.

Finally, the Internet Phone Wizard has two LEDs that indicate what type of call you're engaged in: a Skype call or a regular phone line call. It gets its power from the USB port, so that's one less power adapter to worry about,

Figure 3-17. Skype's contextual menu with the Internet Phone Wizard installed

too. For more information about the Internet Phone Wizard, see its manu-facturer's web site at *http://www.actiontec.com/*. To see how you can use the Internet Phone Wizard to provide Skype network access to an Asterisk PBX system, check out "Connect Asterisk to the Skype Network" [Hack #98].

CHAPTER FOUR

Asterisk
Hacks 41–58

The Linux domain of free software is a land flowing like milk and honey with telephony hacks—the hackers' Promised Land, so to speak. Of course, many of these hacks translate to BSD, and even to Mac OS X, since they're cast from a similar Unix mold.

In this chapter, I'll cover Asterisk—the open source telephony server designed originally for Linux, but now available for Mac OS X and BSD. Asterisk is a workhorse, a flexible, open system that's the telephony equivalent of Apache, the world's most widespread web server.

Because of its modularity and flexibility, Asterisk is as much a platform as is Linux. It's sort of become the cornerstone of Linux-based telephony, thanks to a vibrant developer community and a sound, open source foundation.

Getting Telephony Devices Connected to Asterisk

Besides implementing "pure VoIP"—voice calls over packet networks like the Internet or your Internet Protocol local area network (IP LAN)—Asterisk can also handle legacy telephone technologies, such as analog phones and phone lines, T1 lines, and various kinds of legacy signaling methods. A large and growing selection of PC expansion (PCI) cards are available that facilitate connecting analog phones and phone lines to an Asterisk server. So, if you want, you can build an Asterisk server that doesn't use VoIP at all—just legacy technologies like analog phones. Or you can build a server that bridges those previous-generation devices with Voice over IP.

The Asterisk software is maintained by Digium, Inc. (*http://www.digium. com/*), a manufacturer of many of the interface cards (and VoIP gateway devices) Asterisk supports. You can certainly use other interface cards with Asterisk, such as those manufactured by Sangoma and VoiceTronix. These manufacturers provide drivers for Asterisk's Zaptel driver framework that allow Asterisk to use them.

To FXO or to FXS, That Is the Question

To use traditional analog telephones and lines with an Asterisk server, you'll need to understand the difference between FXO and FXS. Their definitions are a source of some confusion, even among telecom folks. FXS (foreign exchange station) interfaces are used to connect telephones, which are FXS devices. FXS interfaces cause the Asterisk server to appear like the telephone company's central office switch when you plug in a phone. FXO (foreign exchange office) interfaces, on the other hand, are used to make your Asterisk server appear like a telephone so that you can connect it to the central office switch.

So, FXO interfaces connect your server to the phone company, and FXS interfaces connect your server to analog phones. Keep this distinction in mind as you work through the hacks in this chapter.

FXS and FXO interfaces are manufactured by many companies, including Intel, Digium, Sangoma, and Cisco, and they come in a variety of hardware flavors, too: PCI cards, rack-mountable enclosures, and tiny little single-line "converter boxes" that are reminiscent of Ethernet media converters. Though these interfaces are self-contained, standalone devices, they tend to be called *media gateways*, or just *gateways*.

> For IP phones and Internet-based connections to the phone company, there is no FXO/FXS vernacular and no legacy signaling involved at all. When there's no legacy signaling, it's called *pure VoIP*.

And Then There Was T1

Digital circuits that employ the T1* carrier (the most widespread type of digital telecommunications connectivity) can also be used to connect legacy phones to the Asterisk server, and to connect the Asterisk server to central office switches. Depending upon your needs and on what is available from your phone service provider, you might employ a primary rate interface (PRI) to hook up to 23 phone lines at a time to an Asterisk PBX, all on a single T1. Likewise, you can connect a T1 to an FXO interface box (a *media gateway*) to connect analog phones to the server, or you can connect a T1 to

---

* In Europe, T1s are called E1s, use a different voice codec, and have 30 phones lines.

a device called a *channel bank* to connect 24 legacy analog phones or analog phone lines.

I've chosen hacks that will let you experiment with Asterisk while avoiding the relatively high cost and management overhead of T1s, though. I don't have a T1 in my home or in my business test lab, and I don't expect you to, either. Fortunately, though, lots of other great sources of information about T1 are available. For starters, check out *T1: A Survival Guide* (O'Reilly). Then, when you're ready to integrate legacy digital telecom into Asterisk, check out *Switching to VoIP* (O'Reilly) and *Asterisk: The Future of Telephony* (O'Reilly) for the details. You'll also get a much deeper exploration of Asterisk and enterprise telephony, to boot!

Now, let's get hacking, shall we?

Turn Your Linux Box into a PBX

HACK #41

Install and test the Asterisk open source telephony server on your Linux PC.

Some RPM packages are available to simplify Asterisk's installation, but manual compilation is relatively easy. So I'm going to show you how to download, compile, and install Asterisk the "old-fashioned" way. The development branch you'll download from is stable, though once you get comfortable with Asterisk, you'll want to jump out on the bleeding edge and try the developer releases, too. Each release tends to introduce something new and worthwhile, even if it's not in the stable branch yet.

The easiest place to download the Asterisk software is the CVS repository at Digium, the company responsible for Asterisk and some of the hardware components that work with it. To access the CVS repository, you'll need to be logged into your Linux computer at a shell prompt as root. Type these commands to run the CVS check-out routine and download the source code:

```
$ cd /usr/src
$ export CVSROOT=:pserver:anoncvs@cvs.digium.com:/usr/cvsroot
$ cvs login
$ cvs checkout zaptel libpri asterisk
```

Alternatively, you can specify a particular version of Asterisk:

```
$ cvs checkout -r v1-2 zaptel libpri asterisk
```

When prompted, use **anoncvs** as a password. If you don't use */usr/src* as the local location for compiling programs, substitute the appropriate path. The CVS client you're running here will create the */usr/src/asterisk* directory that contains all the Asterisk source code. Once the download completes, you are ready to begin compiling.

Asterisk consists of several software components for Linux. Not all of these packages are required, as some of them are drivers for Digium's interface cards. If you aren't planning to use Digium's cards, you'll need to build only the last of the three, *asterisk*:

libpri

> A driver module that supports Zaptel-compliant interface cards (described in this chapter's introduction) so that ISDN and PRI trunks can be interfaced with Asterisk

zaptel

> A driver module that allows legacy telephone line interfaces cards that provide FXO, FXS, and T1/E1 signaling to be used with Asterisk

asterisk

> A modular software daemon that provides telephony, management, and call-accounting features, including voicemail, Session Initiation Protocol (SIP) telephone support, dial plan, and so on; in a nutshell, Asterisk is an all-software PBX

If you're wondering about these technical terms, don't worry. As you experiment with Asterisk and learn more about VoIP, they'll become very familiar. For now, just compile and install all three packages.

After you run the CVS download, the source code for each Asterisk software component is sitting in its respective directory in */usr/src*. Let's compile each software component by issuing the following commands. Again, you need to compile *zaptel* and *libpri* only if you're planning on using legacy or Digium interface cards. Many of the examples in this book use legacy devices, so it's probably a good idea to compile them all right now. Here is the sequence of commands:

```
# cd zaptel
# make clean ; make install
# cd ../libpri
# make clean ; make install
# cd ../asterisk
# make clean ; make install
```

 Do compile Zaptel before you compile Asterisk, or else Zaptel features will be missing from the Asterisk build. What is Zaptel, you ask? Keep reading....

It should take 20 minutes at most to complete the whole build on an average PC. Once built, Asterisk is ready to use. But you can't race a Ferrari without a training lap on the test track, and you can't really use Asterisk until you understand the basics of configuring it. So it's time for driving school. To get started, run this command in the Asterisk source directory:

```
# make samples
```

This creates a basic sample set of Asterisk configuration files and places them in */etc/asterisk*. You might want to peruse these files—especially *extensions.conf* and *sip.conf*, where you'll likely be spending a lot of time.

If you've used an RPM package or some other precompiled Asterisk distribution (or if you've obtained a Linux distribution with Asterisk already installed), you can still obtain the source distribution files from Digium's CVS repository and issue only the make samples command. This will give you the sample configuration files without actually rebuilding Asterisk on your PC.

Start and Stop the Asterisk Server

The Asterisk program has two modes of operation: *server* mode and *client* mode. The server is the instance of Asterisk that stays running all the time, handling calls, recording voicemails, greeting callers while users are away, and so on. The client is the instance of Asterisk that allows you to monitor and manipulate the server while it runs. The mode the program uses depends on how Asterisk is invoked at the command prompt or within a shell script.

To launch Asterisk in server mode, execute this command:

```
# asterisk -vvv &
```

The more *v*'s, the more verbose Asterisk's console output will be.

To connect Asterisk in client mode on the local machine already running in server mode, execute this command:

```
# asterisk -r
```

Once the Asterisk client is connected to the Asterisk server, you can use Asterisk's command-line interface to issue queries and commands about the telephony server. These include listing calls in progress, listing used and unused channels, and stopping the Asterisk server.

You can shut down the server using one of several Asterisk CLI commands:

- Restart now
- Restart when convenient
- Stop now
- Stop when convenient

The "restart" commands stop and then restart the Asterisk server process, which can be helpful in situations where the server's configuration has changed significantly and needs to be restarted. The "stop" commands just shut down the Asterisk server process. You'll have to execute the Asterisk program in server mode to get it running again.

The "now" and "when convenient" arguments tell Asterisk how quickly to shut down or restart. If you want to interrupt the current calls and tasks in progress on the server, "now" is appropriate. If you want Asterisk to wait until all the calls and tasks are finished and there is no call activity at all, "when convenient" is appropriate. Generally, especially if you're planning to have any callers besides yourself on the system, get in the habit of using "when convenient."

> All of these commands ultimately shut down Asterisk. If you make a configuration change that doesn't require a complete restart, like a change to a certain phone extension, you can just use reload at the Asterisk prompt.

Linux-Specific Start and Stop Scripts

Depending on your particular flavor of Linux, be it Fedora, Debian, SuSE, or something else, you'll find your system's normal startup scripts in a place that's unique to each flavor. Fortunately, Asterisk's *Makefile* has an option that lets you automatically generate start and stop scripts that are specific to your flavor of Linux. In your Asterisk *src* directory, just issue the command make config, and the scripts will be installed. These scripts start and stop not only Asterisk, but also the Zaptel drivers, if you've compiled them.

As it stands at this point, your Asterisk server won't be especially useful. You'll be able to explore the Asterisk command prompt with asterisk -r, but the truly fun stuff, like hooking up phone lines and phones, is still to come. To try out Asterisk's cool demonstration routines—like interactive voice response (IVR) and an Internet-based VoIP call—you've got to configure a phone of one sort or another to access the Asterisk server. Keep reading!

HACK #42 Attach a SIP Phone to Asterisk

Asterisk is a phone system. But it won't do you much good without some phones connected.

You're about to use a SIP telephone to access the de facto auto-attendant greeting and to access a brief demonstration of an Inter-Asterisk Exchange (IAX) trunk over the Internet. Sound like too much? Don't worry; most of this is already configured with Asterisk out of the box. The toughest part for a VoIP beginner will be making sure Asterisk is willing to answer SIP calls—and that's pretty easy.

You won't need a regular phone line for this hack—just a SIP phone, Asterisk, and an Internet connection.

SIP is one of several standards that allow IP voice endpoints and application servers such as Asterisk to establish, monitor, and tear down media sessions across the network. Asterisk uses SIP to facilitate calls on behalf of SIP-based IP phones such as the BudgeTone 101, the Cisco SIP IP Phone 7960, and the Avaya 4602. I've chosen the BudgeTone 101 hardware because it's cheap, but you can go even cheaper and apply this hack using a softphone like the X-Lite [Hack #4], which is free.

The Sipura SPA-841 is another excellent low-cost SIP phone.

Configure a Grandstream BudgeTone 101 IP Phone

The BudgeTone 101 phone has a Menu key, an LCD display, and two arrow keys that you use to navigate its configuration menu options: DHCP, IP Address, Subnet Mask, Router Address, DNS Server Address, TFTP Server Address, Codec Selection Order, SIP Server Address, and Firmware Versions (called *Code Rel* on the phone's screen). When you get to the option you want, you press the Menu key to select it, and then you enter the numeric data required for each option using the keypad. Use this menu only to set up the IP address, subnet mask, and router (default gateway) address.

To get the phone enabled for the next configuration step, turn off DHCP and assign an IP address, subnet mask, and router address.

More advanced configuration is performed using the BudgeTone's built-in web configuration tool. When you access the IP address you assigned to the phone using your web browser, you'll be prompted to log in to the phone. The default password is *admin*.

Then, you'll be confronted with a big page of configuration options. Many of these options are available only through this interface, not from the phone's keypad menu. After you apply your configuration changes, you need to power-cycle the BudgeTone.

Some IP phones offer a Telnet interface rather than (or in addition to) a web-based one. To use these tools, you must connect to the phone with a Telnet client rather than with a web browser. In any event, once you've set the network configuration on the BudgeTone, ping its address from another host on the same network subnet to make sure it's speaking Transmission Control Protocol/Internet Protocol (TCP/IP).

Set the IP Phone to Use a SIP Server

The IP phone, whose address I'll assume is 10.1.1.103, must be set to use your Asterisk box as a SIP server if you're to interact with the Asterisk demo. In your test lab, the IP phone should refer to the IP address of the Asterisk server (10.1.1.10, say) being used as its SIP server. Configure the SIP User ID setting as 103, too. For the DTMF Mode option, select SIP Info. Then apply the config changes and reboot the IP phone. (The same configuration options are supported by other makes of SIP phone, too.) The configuration page for a BudgeTone phone that has been configured to use a local SIP server (your Asterisk box) is shown in Figure 4-1.

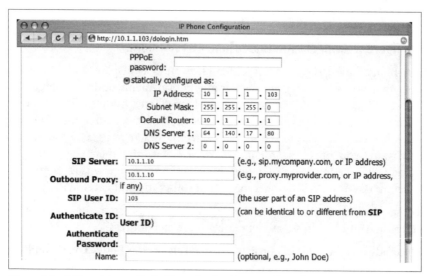

Figure 4-1. A Grandstream BudgeTone that has been configured to use a SIP server at 10.1.1.10

Allow the IP Phone to Place Calls Via Asterisk

Until you authorize a SIP phone to communicate with Asterisk using Asterisk's SIP configuration file, you will always receive SIP error messages when trying to dial to (or through) the Asterisk server. This is Asterisk's way of ignoring what it sees as an unauthorized endpoint. Unlike traditional PBXs, which tend to give network access to any phone connected on an active port, SIP servers tend to enforce some security—usually in the form of password authentication.

So tell the Asterisk server to stop ignoring requests from your IP phone. Asterisk, the softPBX, refers to IP phones and other SIP devices as *channels*. SIP channels (or peers, if you like) are defined in Asterisk's configuration file, */etc/asterisk/sip.conf*. To enable the phone as configured in Figure 4-1, add the following to the end of this file:

```
[defaultsip]
type=friend
context=default
username=103
fromuser=SIP Phone
callerid=103
host=10.1.1.103
nat=no
canreinvite=yes
dtfmode=info
disallow=all
allow=ulaw
```

The preceding configuration settings add the 10.1.1.103 IP phone that matches the configuration of the Grandstream BudgeTone. Take note of the username, callerid, and host values, which resemble each other (103) in this case. They don't need to resemble each other, however, because there's no relation between a phone IP address and its SIP username or caller ID. These can all be completely different.

One of the biggest differences between SIP and its predecessor, H.323, is that SIP identifies its phone endpoints (or *terminals* in H.323-speak) by IP address (and port number) exclusively, whereas H.323 still relies on their Ethernet MAC hardware addresses. This makes SIP more flexible!

There are two ways to enable the configuration change you've just made. One is to restart Asterisk:

```
# asterisk -rx restart
```

Bear in mind that restarting your softPBX might be acceptable at home or even in a small office environment, but you'd better make sure no calls are in progress if you restart it in any production environment, lest you draw the ire of angry phone users. Perhaps a better way to handle the addition of a new endpoint to the softPBX is the reload method. To do this, issue the Asterisk reload command using the -rx shell option:

```
# asterisk -rx reload
```

Or log into the Asterisk CLI (as in "Turn Your Linux Box into a PBX" **[Hack #41]**) and issue the Asterisk reload command:

```
pbx*CLI> reload
```

No calls are interrupted when the reload occurs. This should keep everyone who is using the system at that moment happy.

Now, you can place calls to the Asterisk server and to the other peers and channels that will be connected to it. The default configuration installed with Asterisk when you compiled it allows for several interesting demonstrations of its capabilities using a SIP phone. (You also can try them using an analog phone, if you have a Zaptel card installed and a phone hooked up—but that's another project **[Hack #44]**.)

Listening to Asterisk

In its default configuration, Asterisk has an auto-attendant that can route calls. To try it out, take the IP phone off the hook and dial 2. Then dial the BudgeTone's Send button. You will hear a friendly voice saying, "Asterisk is an open source, fully featured PBX and IVR platform...."

Try this demo while watching the call progress on Asterisk's console by issuing asterisk –vvvvvr at a Unix shell before beginning the call.

While listening to the automated attendant greeting, dial 500. This will cause the Asterisk server to greet you, connect you to a server at Digium, Inc. using the Internet, and allow you to listen to another automated greeting—the one being played back by a production Asterisk PBX at Digium's office. This connection does not use the Public Switched Telephone Network (PSTN) at all, but rather, a Voice over IP "trunk" that is set up on the fly by Asterisk.

The Voice over Internet demo requires User Datagram Protocol (UDP) port 4569. If you're using a firewall or NAT device, be sure it permits outbound traffic on this port. Most broadband routers will permit this type of traffic by default.

You can also perform an echo test by dialing 600, and you can access Asterisk's built-in voicemail service by dialing 8500. This will give you at least some idea of how your voice sounds when it's been processed and played back for the person on the other end of a call.

Connect a Phone Line Using an FXO Gateway

HACK #43

The easiest way to interface Asterisk to a standard phone line—like the traditional phone line in most homes—is by using a telephone media gateway.

There are essentially two ways to connect a traditional, non-IP phone device—be it an analog phone, a digital phone interface, or a telephone line—to the Asterisk system. The first way is via the Zaptel telephony framework, a driver standard that permits telephony interfaces to be used with Asterisk. (Oddly, Zaptel's inventor, Jim Dixon, named his creation after the early 20th century Mexican revolutionary, Emilano Zapata.)

Zaptel engineering is a very deep subject that warrants its own hack [Hack #44], and even its own book (such as O'Reilly's *Asterisk: The Future of Telephony*), so we're going to start with the *other* way of connecting non-IP phone devices to Asterisk: via media gateways. Once you've got this down, you can move on to the wonderful world of Zaptel.

A media gateway is a device that offloads the responsibility of hardware interfacing from the server. It converts non-IP signaling into VoIP signaling and vice versa. Media gateways don't need driver frameworks like Zaptel to support connecting phone lines or other legacy technology. They come ready to install on the network, with no software to compile. Just plug, configure, and go. Connecting legacy phones and phone lines to Asterisk via a media gateway is decidedly easier than using Zaptel, so that's the route we're taking here.

The analog telephone adapter (ATA) pictured in Figure 1-1 of "Get Connected" [Hack #1] is a media gateway, since its job is to connect an analog phone to a VoIP service provider's SIP server over the Internet. In this case, though, we want to do the reverse of that. We want to connect a telephone company line to our own server. We'll do it using a Clipcomm CG-200 media gateway, an inexpensive, Korean-made gateway that supports connecting up to two phone lines (the CG-400 supports up to four).

When complete, the phone line, connected to the media gateway, will be answered automatically by the Asterisk server, and a greeting message will be played for the caller.

Configure the Gateway

The Clipcomm CG-200 and CG-400 are similar to other phone media gateways: they provide an Ethernet interface (or two) and two or more FXO ports that each allow you to connect a telephone line. But unlike some other gateway hardware, the Clipcomm has a fantastic web-based configuration interface, as shown in Figure 4-2.

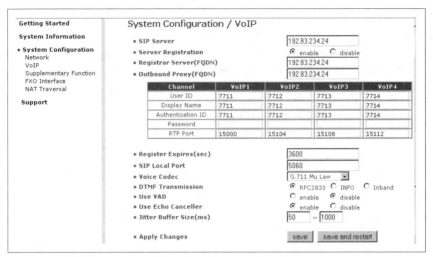

Figure 4-2. The Clipcomm's web-based configuration interface

Let's assume you've already set up the TCP/IP basics on your Asterisk server machine and on your media gateway. Make sure you can ping back and forth between them, too. Then, accessing the VoIP configuration page on the Clipcomm, enter the Asterisk server's IP address into the SIP Server field. Enable SIP Registration and enter the Asterisk server's IP address again into the Registrar and Outbound Proxy fields. This will cause the media gateway's SIP client to register with the Asterisk server.

> *Registration* is the process by which a SIP client is authenticated with the SIP server and is also the means by which the SIP server knows how to reach the endpoint in case it needs to route a call to it. Just as a TCP/IP device can register with a DHCP server, a SIP client can register with a SIP server, called a *registrar*.

Now, you'll need to tell the media gateway what credentials (username and password) to use for each phone line you're going to be "passing through" to the Asterisk server. If you're just connecting a single line, you need only establish credentials for VoIP1. (VoIP2, VoIP3, and VoIP4 correspond to the second, third, and fourth phone lines you can connect.)

The User ID tends to be a phone number. When this User ID is dialed by a caller, it signifies that this SIP endpoint should be called. This differs from the Authentication ID, which is used to register the SIP endpoint with the Asterisk server. Authentication ID and User ID needn't be the same, but they often are. In this case, we've chosen 7711 for both. This is the number we'll use later with Asterisk to handle calls to and from the phone line that's connected to this media gateway.

Click the Save and Restart command button to reboot the media gateway. Then click Supplementary Function. This will pull up a page similar to the one in Figure 4-3.

VoIP Supplementary Function FXO Interface NAT Traversal **upport**					
Channel		**VoIP1**	**VoIP2**	**VoIP3**	**VoIP4**
1-Stage FXO Gateway		○	○	○	○
2-Stage FXO Gateway		○	○	○	○
Use FXO PIN Code		☐	☐	☐	☐
FXO PIN Code(4 digits)		0000	0000	0000	0000
Call Forwarding to VoIP		○	○	○	○
Call Forwarding to PSTN		◉	◉	◉	◉
Forwarding Condition		Unconditional	Unconditional	Unconditional	Unconditional
Call Forwarding Number					

Channel		**PSTN1**	**PSTN2**	**PSTN3**	**PSTN4**
2-Stage FXO Gateway		○	○	○	○
Use FXO PIN Code		☐	☐	☐	☐
FXO PIN Code(4 digits)		0000	0000	0000	0000
Call Forwarding to VoIP		◉	◉	◉	◉
Forwarding Condition		Unconditional	Unconditional	Unconditional	Unconditional
Call Forwarding Number		7711	7712	7713	7714

• End of Call on PSTN Digits   ○ enable   digits [*#*]
                                     ◉ disable

Figure 4-3. The Clipcomm's supplementary function configuration page

On this page, you can see the FXO channels that correspond to the VoIP channels. They're called PSTN1, PSTN2, etc., and they represent the two or four RJ11 jacks on the gateway's back panel. To get incoming calls from the attached phone line to be forwarded automatically to the SIP server on the Asterisk machine, click the radio button under PSTN1, labeled Call Forwarding to VoIP. Now, when calls come into the Clipcomm from the phone-company line, they'll be answered automatically, and the Clipcomm will attempt to route them through to the Asterisk server. Don't forget to save this configuration.

> If we wanted SIP calls outgoing from Asterisk *to* the Clipcomm to be forwarded to the PSTN, we would need to enable Call Forwarding to PSTN for each VoIP channel, too. Keep this in mind for the next hack.

Configure an Asterisk SIP Peer for the Gateway

Asterisk refers to SIP endpoints as SIP peers, and it uses */etc/asterisk/sip.conf* to establish settings for them: everything from usernames and passwords to basic audio preferences. Our objective here is to establish a SIP peer configuration for the media gateway we've just configured. So, using your favorite text editor, add the following to your *sip.conf* file:

```
[7711]
callerid="Outside Line" <200>
canreinvite=no
context=default
dtmfmode=rfc2833
host=dynamic
port=5060
type=friend
username=7711
```

Notice how the bracketed heading and the username setting of 7711 match the media gateway's User ID and Authentication ID settings, respectively.

Make Asterisk Answer Automatically

Now, save *sip.conf* and open up */etc/asterisk/extensions.conf*. This file tells Asterisk what to do whenever a user dials a phone number. It contains the "dial plan" that guides the system-wide call-handling functionality of the Asterisk server. In the [default] section of the file, comment everything out and add these lines:

```
exten => s,1,Answer
exten => s,2,Playback(abandon-all-hope)
exten => s,3,Hangup
```

Save the file and launch (or relaunch) Asterisk:

```
# asterisk -rx reload
```

Now, calling the phone number of the line connected to the media gateway will result in the call being answered by the Asterisk server. You'll hear a voice message, and then the call will be hung up.

Connect a Legacy Phone Line Using Zaptel

HACK #44

You don't need to buy a VoIP phone to make use of Asterisk—use your home phone instead.

About a dozen interface cards support the Zaptel standard, allowing you to connect something as simple as a two-wire analog telephone, or something as sophisticated as a digital T1 breakout box (called a *channel bank*) for connecting digital business phones. In this hack, you'll see how to connect a phone line using such an interface card—either an X100P or a TDM400P, both manufactured by Digium.

To build a Linux PBX that can communicate with the PSTN (the network that 90 percent of the world still uses for telephony), you'll need at least one trunk channel to communicate with the FXS interface in the phone company's central office switch. This channel will provide you with a dial tone from your local phone company, so calls to and from the PSTN can be handled by the Asterisk server. There's not much to setting up an FXO channel with Asterisk. One way, covered in this hack, is to install an X100P, TDM400P, or similar FXO line card in the Asterisk server. (The other way is to use an FXO media gateway [Hack #43]).

> The connection from your premises to the phone company's switch is called the *local loop*.

Install an Interface Card

To get started, you'll need to obtain an Intel i537-based FXO interface card such as the Digium X100P. (If you'd like to save a few bucks and build your own X100P clone, check out "Brew Your Own Zaptel Interface Card" [Hack #64].) Install the X100P card into your PC's PCI bus (sorry Mac users, you're stuck using a media gateway, as covered in the previous hack) and connect an RJ11 standard phone patch cord from the wall jack of an active telephone company Plain Old Telephone Service (POTS) line into the appropriate port of the X100P. On the X100P, this port is the one marked with an etching of a telephone wall jack.

Now, download and compile the Zaptel driver and Asterisk [Hack #41]. This creates the zaptel and wctdm modules, which need to be loaded during startup, by adding this code to the script that launches Asterisk, right before the line where Asterisk itself is launched:

```
modprobe zaptel
modprobe wctdm
/etc/rc.d/init.d/asterisk start
```

By now, the card is in a PCI slot on the Asterisk server, the phone line is connected, and you've compiled and installed the Zaptel drivers. Your next step is to define the FXO trunk connection as a channel that is usable by Asterisk. Once defined, you can reference the channel within your Asterisk call-routing scheme. The POTS line can serve as the full-time gateway for all PSTN calls and all telephones in your home or office. Or the POTS line can just be a connection mechanism so that the Asterisk server can answer incoming calls on the POTS line if they aren't answered by a person within a certain number of rings. But first, the FXO connection must become a named Asterisk channel.

Each voice channel in Asterisk has a number. This number consistently represents the same channel throughout all of Asterisk's configuration files and in its logging output. The numbering of voice channels—especially those that require a dedicated piece of interface hardware in the server—is determined by the order in which their drivers are loaded and the order in which they are identified in the PC's PCI bus. Figuring out which card is which—say, in a situation where you have just installed three or four X100P cards, each with its own POTS line—can require a bit of trial and error. In this project, we're using only one card and one line, so it should be a breeze.

The voice channel we're going to create will be called Zap/1-1. Asterisk follows a similar convention when naming all voice channels, even if they aren't analog phone line channels. The channel name is divided into two pieces. The first piece, Zap/1, refers to the physical Zaptel interface channel (which is either an FXO/FXS interface or a PRI channel). The second piece, -1, refers to the line number (more on multiple-line interfaces later).

> For Zaptel interfaces that support only a single line, you can refer to them without the line number—i.e., Zap/1 and not Zap/1-1.

Assuming you haven't touched the Asterisk configuration files since running make samples in the first Asterisk hack, you'll have to make only two quick config changes to fire up your POTS line. The first change is in */etc/zaptel.conf*. Add the following lines to the end of the file:

```
fxsks=1
loadzone=us
defaultzone=us
```

The first line tells the Zaptel configuration program, ztcfg, to set the X100P card to use FXS *Kewlstart* signaling—a variation of conventional FXS loop-start signaling. The number 1 is referenced because only one Digium card is installed, and it has only one channel, like the X100P card, so it's card number one and its channel will be 1 as well. If two cards were installed side by side, the first line would say fxsks=1-2 instead. If there were more than one channel per card (like the TDM400P), a single channel number would be used for each channel on that card—i.e., fxsks=1-4 for a card with four lines attached. fxsks=1-8 would work fine if you had two TDM400Ps installed with four FXS modules apiece. The next two lines in the code snippet localize the

FXS signaling functionality of the X100P interface with loadzone and defaultzone. Other valid zones include fr, de, and uk.

Now, you might be asking yourself, "Why am I configuring the FXO interface card to use FXS signaling?" The answer is simple: to communicate with the FXS device interface at the central office, the local interface must use FXS signaling. Recall from this chapter's introduction that only FXS devices can receive signals meaningfully from FXO devices, and vice versa.

> To alleviate confusion over FXS/FXO kernel module naming, wcfxs has been deprecated in favor of wctdm in releases of Asterisk later than 1.0.5.

The next change you need to make is in the */etc/asterisk/zapata.conf* file. The sample configuration should be completely commented out (comments are denoted by semicolons at the beginning of the line). If it's not commented out, place a semicolon at the front of each line. Then, add the following lines to the end of the file:

```
context=default
signalling=fxs_ks
usecallerid=yes
echocancel=yes
callgroup=1
pickupgroup=1
immediate=no
channel=>1
```

The first line tells Asterisk what set of assumptions to make (i.e., what "context" to choose) when handling calls coming in on the POTS line. The second line tells Asterisk (not ztcfg) what type of signaling the X100P has been set to use. The following lines turn on a few traditional telephony features—caller ID, echo cancellation, and other stuff that's covered in more detail later. The last line assigns all the previous settings to channel 1. The assignment of these inherited settings uses the => assignment operator rather than just an equals sign (=). The Asterisk configuration parser doesn't distinguish between them; the convention is merely for ease of human readability.

> *Contexts* are Asterisk's way of meaningfully grouping call-flow scenarios. A context describes what behavior is caused by dialing 1 at the outset of a call, while another context describes what behavior is caused by dialing 1 at some point thereafter. These contextual behaviors are defined in */etc/asterisk/extensions.conf*.

You'll need to make one more quick change to Asterisk's sample configs: change the Zap/g2 definition for $TRUNK in *extensions.conf* to Zap/1. (This step might not be necessary with earlier versions of the sample config.) This will allow outbound dialing to be directed to the correct channel, Zap/1, the one that represents the connection to the PSTN.

Now, since you've added a new hardware interface, you must restart Asterisk. Once you've done that, try calling the POTS line you've connected to the X100P using a second phone line or your cell phone. After a few rings, assuming you haven't changed the configuration, Asterisk will answer and you should hear the familiar demo greeting that you heard in "Turn Your Linux Box into a PBX" [Hack #41]. If you examine Asterisk's console output during this demo, you'll see something like this:

```
-- Starting simple switch on 'Zap/1-1'
-- Executing Wait("Zap/1-1", "1") in new stack
-- Executing Answer("Zap/1-1", "1") in new stack
...
-- Playing 'demo-abouttotry'
-- Executing Dial("Zap/1-1", IAX@/guest@misery.digium.com/s@default) in new
stack
...
```

Through the console output, you can trace every step Asterisk took to recognize, answer, and process the incoming analog call from the PSTN and to connect it using the IAX protocol to a remote server across the Internet. Note that although this chapter is about legacy, circuit-switched telephony, we're using IAX to get our feet wet with VoIP. Plus, the IAX demo is so easy to run with Asterisk "out of the box" (it isn't broken by broadband routers the way SIP often is) that it's a great way to demonstrate how a VoIP signaling protocol can be used with legacy signaling on the PBX.

H A C K   Forward Your Home Calls to Your Cell Phone
#45
Using Asterisk, you can create a simple call forwarder, so calls to your home can follow you whenever you go.

Asterisk is a programmable platform in the same way that the Apache Web Server is. There are many ways to program Asterisk, but all of them connect in some way to the core of Asterisk's functionality—its so-called "dial plan." The dial plan begins (and usually ends) in */etc/asterisk/extensions.conf*. Using the dial plan, you can program how your softPBX should behave, which phones should ring when different digits are dialed, how long they should ring, and what to do if nobody answers when they ring.

So, it's actually pretty straightforward to program the dial plan to forward all incoming calls from a certain line to another phone number via a second line. There are many uses for this, including having your phone calls follow you wherever you go, as well as using the Asterisk server as a screen so that you don't have to give people the number you are forwarding to. This is a great way to keep your cell phone or parents' home phone number private. The Asterisk server will dial the cell phone on the second line and then bridge (or conference) the two lines together for the duration of the call.

The hardest part about setting up this configuration is connecting two lines (or two SIP peers acting as lines [Hack #43] to the Asterisk server. Once that's done, the forwarding part is simple. But before we get to that, let's check out the configuration for the two lines.

Let's assume that two SIP peers are connected to the Asterisk server, vis-à-vis a media gateway with two SIP clients, like the Clipcomm used earlier [Hack #43]. The *sip.conf* configuration for these peers would look something like this:

```
[7711]
callerid="Outside Line 1" <200>
canreinvite=no
context=incoming
dtmfmode=rfc2833
host=dynamic
port=5060
type=friend
username=7711

[7712]
callerid="Outside Line 2" <201>
canreinvite=no
context=default
dtmfmode=rfc2833
host=dynamic
port=5060
type=friend
username=7712
```

Line 1 is SIP peer 7711, and line 2 is SIP peer 7712. Let's say that the line we're going to receive calls on is line 1, and the line we're going to use to call the cell phone is line 2. Note context=incoming. This creates a context within Asterisk for incoming calls to arrive. Now, open up *extensions.conf* so that you can create a dial plan—a set of instructions that tell Asterisk what to do in this incoming context—to correspond to peer 7711's incoming context setting:

```
[incoming]
exten => s,1,Dial(SIP/7712/${CELL_PHONE},30)
exten => s,2,Playback(abandon-all-hope)
exten => s,3,Hangup
```

Since *sip.conf* indicates that all incoming calls from SIP peer 7711 should enter the incoming context, we've created that context in the dial plan (as shown in the previous code snippet). Using the special s extension, whose purpose is to incorporate into the context incoming calls that haven't been triggered by a user dialing an extension number (calls like those incoming from the outside world), we can specify three steps to deal with the call:

1. Dial the cell phone number on the second SIP peer (7712). You can specify the number—for instance, 1-440-864-8604—instead of using the ${CELL_PHONE} variable, since we haven't really covered variables yet. The 30 specified in the first command says to attempt to bridge (conference) the call on the two lines for up to 30 seconds before giving up.

2. If the call isn't bridged because the Dial command times out, the Playback command will play a greeting. In this case, I've specified a greeting called abandon-all-hope.

3. If either of the previous steps cannot be completed, the caller will be disconnected.

So, when an incoming call from SIP peer 7711 hits the server, SIP peer 7712 (the "second line") will be directed to call your cell phone and attempt to bridge the call.

I know what you're thinking: "Why wouldn't I just use my phone line's built-in call-forwarding service to do this?" My answer is "Selectively Forward Calls" [Hack #46].

HACK #46 Selectively Forward Calls

You can pass caller ID signals into Asterisk, and have them acted on appropriately—including auto-ignoring the people you don't want to speak to.

By making some clever use of Asterisk's built-in caller ID channel variable and a little workflow logic, it's easy to turn your call-forwarding project from the previous hack into something even more useful. In this hack, we'll make Asterisk forward calls to your cell phone only if they're from a certain caller ID. That way, you need only be bothered with answering your cell phone if dear old Mom is calling (or your boss).

Asterisk refers to the one or more voice communication links of a phone call as channels. So, when a call-forwarding setup that uses two SIP peers is active, it's said to use two channels. Each channel has with it a number of channel-specific variables that contain information about the ongoing call. When the call ends, the channels, and these channel-specific variables, disappear. One of these variables is ${CALLERIDNUM}, which contains the phone number of the calling party, as signaled by the calling peer. (On the PSTN, caller ID signals originate from the exchange switch of the calling party.) We can use this variable to figure out whether we want to forward a call.

> Unless you're paying for caller ID service, your Asterisk server won't receive caller ID signals, and this hack won't work. Some phone companies (and just about all VoIP service providers) include caller ID for free.

Consider the following:

```
[incoming]
; Priority 1: Check to see if the call is Mom's home phone.
; If so, go to priority 5; if not, continue to priority 2.

exten => s,1,GotoIf($["${CALLERIDNUM}" = "3138853352"]?5:2)

; Priority 2: See if the call is Mom's cell phone.
; If so, go to priority 5; if not, continue to priority 3.
exten => s,2,GotoIf($["${CALLERIDNUM}" = "3132981848"]?5:3)

; Priority 3 and 4: This call's not Mom, so just drop it.
exten => s,3,Playback(carried-away-by-monkeys)
exten => s,4,Hangup

; Priority 5: Dial my cell phone for 30 seconds to connect Mom.
exten => s,5,Dial(${MYCELLPHONE},30)

; Priority 6 and 7: If not answered in time, drop the call.
exten => s,6,Playback(carried-away-by-monkeys)
exten => s,7,Hangup
```

Note the syntax of the GotoIf command. If you're familiar with logic control structures in programming, the ? should look like a "then" in an if-then workflow statement. A colon (:) separates the then-target from the else-target. The targets correspond to the step numbers in each of the exten directives, of course.

If you think Asterisk dial-plan syntax is atrocious, well, you're right. Don't get too hung up on it now, though. There are some good references out there for Asterisk dial-plan commands, including *http://www.voip-info.org/* and the unforgettable classic, *Switching to VoIP* (O'Reilly). For now, just keep hacking, and you'll get comfy.

Hacking the Hack

With a little modification, you should be able to forward incoming calls to *different* numbers, depending on their caller ID values. Just rearrange the previous example so that each GotoIf numbered target step contains a Dial command with a different phone number—one for Mom, one for Dad, etc. You can even forward calls with no caller ID signals (like those from telemarketers) to a fun destination [Hack #48].

Report Telephone Activity with Excel

With a little help from Microsoft Excel, you can dig into your CDRs, chart your top callers, and create utilization records for the users of your PBX server.

Most commercial softPBX systems provide a detailed logging mechanism for keeping track of when and to whom calls were made and received. Asterisk provides this, too. In */var/log/asterisk/cdr-csv/Master.csv*, a flat text log of all call activity is retained. It's a snap to import this into Excel or your favorite spreadsheet for analysis. You can download the file from your server using FTP, or you can run the following command to email it to you (keep in mind that large logfiles might not work well with this trick):

```
# cat /var/log/asterisk/cdr-csv/Master.csv | mail me@mydomain.com
```

Of course, replace *me@mydomain.com* with your email address. If your Linux server has sendmail or a similar Simple Mail Transfer Protocol (SMTP) agent running (most do), the contents of the file will be emailed to you. You can then copy and paste them into Excel, as shown in Figure 4-4. Place the cursor on column A, row 1 before pasting.

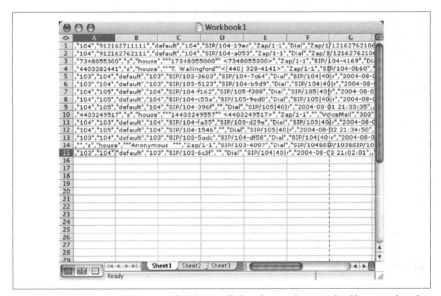

Figure 4-4. A portion of the Asterisk ASCII call-detail-record (CDR) logfile, copied and pasted into the Macintosh version of Excel

Once you paste the text or open the file, select column A by clicking the A column heading. Then use Excel's Text to Columns function, on its Data menu. This will launch a wizard that will help you organize the text file into columns so that it's actually useful within Excel. You'll see a preview of the text you pasted in the bottom portion of the window that appears. Leave the Delimited radio button selected and then click Next.

Select Comma as a delimiting character, make sure no other delimiters are selected, as in Figure 4-5, and click Finish. Now, you're ready to label the column headings according to their purposes. Insert a blank row at the top of the spreadsheet, and you can label them as outlined in Table 4-1.

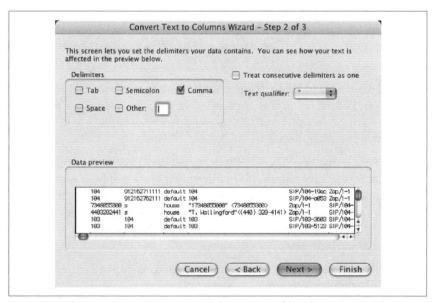

Figure 4-5. The second step of the Text to Columns Wizard breaks up the CDR log text into meaningful cells of data

Table 4-1. Asterisk CDR default fields

CDR field	Description
Account code	A tag that can be used in billing and analysis
Source:	The unique identifier of the endpoint placing the call
Destination:	The unique identifier of the endpoint receiving the call
Context*:	The dial-plan context of the call (more on this later)
Caller ID:	The calling-party identification signals supplied by the calling endpoint
Incoming Channel:	The voice channel that routes to the caller
Destination Channel:	The voice channel that routes to the receiver

Table 4-1. Asterisk CDR default fields (continued)

CDR field	Description
Application*:	The Asterisk software function handling the call
Last Data Sent to Application*:	Information the application uses to connect the call
Start Time:	The time of first contact from the caller to the softPBX
Answered Time:	The time the receiving endpoint answered, if applicable
End Time:	The time of the end of the call, regardless of whether it was answered
Duration:	The length, in seconds, from the first contact to the end time
Billable Duration:	The length, in seconds, of connected, billable time during the call
Disposition:	The last known status of the call during this application
AMA Flags*:	Automated Machine Accounting flags, used by some telephony billing software

The field names marked with an asterisk (*) record Asterisk proprietary information. For example, the Application field might not have a meaningful correlation on another softPBX because not all softPBXs refer to telephony functions as applications.

The idea here is that once the CDR is imported into Excel—or another data-analysis tool—you can interpret it in interesting ways. Suppose you want to figure out which customer places the most calls to your technical support department. You can count occurrences of that customer's caller ID in the CDR. Or, if your teenage daughter is receiving a dozen calls a day, you can bill her accurately for them!

With CDRs in Excel, Crystal Reports, or even a homegrown Perl program, a savvy telephony administrator can do the following:

- Determine which channels are used the most and the least
- Determine which endpoints are called most often
- Calculate the percentage of outgoing calls that are out of your area code
- Create a list of calls, broken down by endpoint
- Create an invoice for a paying subscriber to the softPBX

> Asterisk's CDRs can also be stored in PostgreSQL, MySQL, and even syslog, depending upon the modules you compile and install.

Creating a Call Report

One of Excel's coolest features is the Pivot Table Report. I actually used Excel for years without touching this menu option, passing over it dozens of times, until one day I had to build a sales report for an application I'd been developing. I had a choice between coding the report myself, building it in a tool like Access, or performing the analysis in Excel. The only problem with that last option was that I didn't know *how* to do the analysis in Excel—I knew only that it could be done.

So, I turned to the Pivot Table Report (or should I say, I—ahem—pivoted to it) and built a sales summary in five minutes, which to this day is still in use at the office where I built it. Needless to say, I've sworn by the Pivot Table Report function ever since. I never knew what I was missing out on by passing over that peculiarly intimidating Excel menu option. And, when it comes to those telephony logs, the Pivot Table's uses are many.

By themselves, the CDRs stored by Asterisk (and other soft-based PBX systems) are like any flat file: they've got lots of data but they don't tell you much because they don't provide any analysis. Cue Microsoft Excel. With Excel's Pivot Table Reports, you can generate some very cool call-activity analysis. List your top callers. List your top system users. Or just figure out your total long-distance and local utilization down to the minute to verify your phone bills.

We'll do one report that sums activity (in minutes) by caller and a second report that adds a breakdown of the total minutes for every phone number called by each caller. To get started, we'll first need to get our hands on the Asterisk CDR logfile, as described at the beginning of this hack. Next, we'll insert a blank row at the top of the worksheet and key in the names of the CDR fields at the top of each column, as shown in Figure 4-6. This will be needed to make the Pivot Table Report. (The names of the CDR fields are laid out in Table 4-1.)

Once the CDR columns are labeled, select Data → Pivot Table Report. Now, you'll get a wizard. Click Next on the first step, where you'll find yourself being prompted to provide the name of a data range where the source data exists. In this case, the source data is the first sheet in the workbook—the one that contains the CDR data. Select this sheet and drag-select all of the columns that contain CDR data. Then, return to the wizard window and click Next. The final step asks you where you want to put the report; choose the option to place it in a new worksheet. Then, click Finish. As shown in Figure 4-7, you'll now have a pivot-table toolbar with the names of your CDR columns on it.

Figure 4-6. The CDRs, imported into Excel

Figure 4-7. A pivot-table toolbar with Asterisk CDR fields

Now, you can drag those column names from the pivot-table toolbar to the left and right columns of the blank Pivot Table Report worksheet. Dragging to the left pivot-table column treats the data from that CDR column as a group label. If you're familiar with Crystal Reports or Access, data grouping in reports should be a friendly concept. If not, read on—you're in good hands. Dragging to the right column of the Pivot Table Report worksheet treats the data from the CDR column as summary data.

It's probably easier just to start dragging column headings and see what happens. Start by dragging the Source column to the left column in the pivot table. Next, drag the Duration column to the right column in the pivot table. These two drags will build a report like the one in Figure 4-8, which shows a sum of minutes for each caller on the system over the period of time covered by the CDR worksheet.

In Figure 4-8, the majority of callers are PSTN phone numbers (the 10-digit numbers), though the majority of minutes are from private extensions (104,

Figure 4-8. A Pivot Table Report that shows all the minutes of call activity on the system, broken out by caller

200, etc.). Extension 200 has the most minutes—261. Of course, this report doesn't tell us how extension 200 spent all those minutes (*whom* 200 was talking to), so let's drag another CDR heading from the toolbar to the left-hand column. Drag the Destination column, and the report will now show the minute totals of each phone number *to whom* each caller placed calls, as shown in Figure 4-9.

Experiment with the other columns. What can Excel tell you about your call activity? With these reports, you'll have a handle on precisely who called whom, and when. That way, if your mom ever says, "Why don't you ever call?" you'll have the perfect response: "Mom, we talked 108 minutes last week alone, and 96 minutes the week before!"

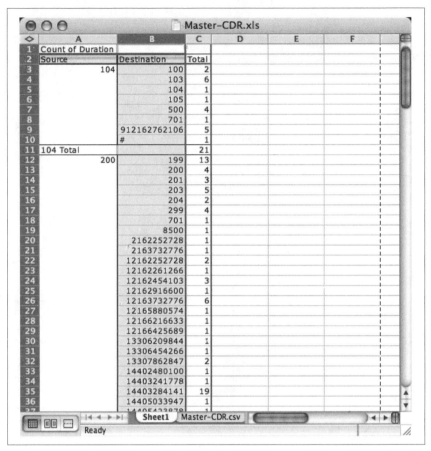

Figure 4-9. A Pivot Table Report that shows detailed minute totals for each extension, for every phone number called

Kindly Introduce Telemarketers to Mr. Privacy

HACK #48

If you enjoy being contacted by anxious telephone pitchmen promising a lower interest rate or offering a great deal on term life insurance while you're just sitting down to dinner, skip this hack.

Using techniques similar to those in "Selectively Forward Calls" [Hack #46], it's possible to discern between phone numbers that supply a caller ID and those that don't. This is different from merely identifying a certain caller ID number and then handling it. What we're doing here is shoveling *all unidentified calls* into a certain action.

If you like, you can even have your phone server handle these calls without interrupting you, putting a decisive end to those annoying dinner-hour calls from "Private" or "Unknown." Using a great little feature in Asterisk, the

PrivacyManager command, we can fight fire with fire. This dial-plan command screens calls as described earlier, identifying the caller ID, or forcing the calling party to enter a caller ID if none is provided at the outset of the call. Best of all, everything can happen without your phone ever ringing, saving you from the aggravation of a sales pitch when you're trying to enjoy a filet mignon.

Consider the following from the [default] context, in *extensions.conf*:

```
exten => s,1,PrivacyManager
exten => s,2,Dial(Zap/2,30)
exten => s,3,Hangup
exten => s,102,Hangup
```

The first priority of this extension contains the PrivacyManager command, which prompts the user to enter his 10-digit telephone number if no caller ID signals have been sent on the channel to identify the caller. If the caller doesn't enter his phone number, he gets dumped to priority 102 (100 plus the current priority), where the call is disconnected using the Hangup command.

 When telemarketers call you, pretend you're Scarface, brandishing your Privacy Manager and saying, "Say hello to my little friend."

If the caller does successfully enter his 10-digit phone number, the dial plan proceeds to the next priority. In the previous example, a Dial command rings a phone connected to a Zaptel card (that's what's referenced by Zap/2) for 30 seconds before giving up and disconnecting the call if nobody answers.

Hacking the Hack

You can combine this hack and "Selectively Forward Calls" **[Hack #46]** to maintain privacy, and to pick and choose functionality based on caller ID—for instance, forwarding calls based on who's calling. And as shown here, you can make sure you *know* who's calling, with your new friend, Mr. Privacy:

```
[incoming]
exten => s,1,PrivacyManager
exten => s,2,GotoIf($["${CALLERIDNUM}" = "3138853352"]?6:3)
exten => s,3,GotoIf($["${CALLERIDNUM}" = "3132981848"]?6:4)
exten => s,4,Playback(carried-away-by-monkeys)
exten => s,5,Hangup
exten => s,6,Dial(${MYCELLPHONE},30)
exten => s,7,Playback(carried-away-by-monkeys)
exten => s,8,Hangup
exten => s,102,Playback(carried-away-by-monkeys)
exten => s,103,Hangup
```

Build a Four-Line Phone Server
#49   Create a simple, fully functional small-office PBX.

An Asterisk server can be the nerve center of two kinds of telephony networks: an all-VoIP network that uses only IP-based connections to route calls, or a hybrid VoIP/legacy network that uses both IP and *time division multiplexing* (TDM) technologies to route calls.

In this hack, you'll use a Digium TDM400P card to turn your Asterisk server into a full-blown PBX that can support up to four legacy phones (or four legacy phone lines) at a time. These legacy devices will be able to call, and be called by, VoIP phones. This setup is depicted in Figure 4-10.

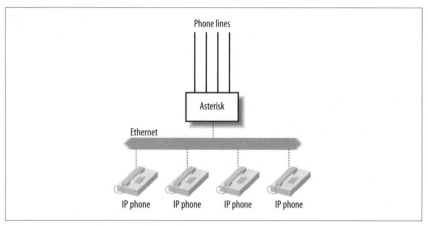

Figure 4-10. Asterisk as a simple small-office PBX

The TDM400P card has four modular interfaces that can host either FXO or FXS modules. FXO modules allow you to connect phone lines to your server, and FXS modules let you connect phones. You can use any combination of FXO and FXS modules (up to four) on a single TDM400P, so you can connect two analog phones and two phone lines, or one phone line and three analog phones, and so on. When you purchase the TDM400P card, you can specify what combination of interfaces you'd like. Here's the main thing to remember, so you don't get the wrong configuration: the red-colored FXO modules connect phone lines, and the green-colored FXS modules connect phones. So, if you want three analog phones to share a single analog phone line, you would use three FXS modules and one FXO module.

The Linux driver framework that allows Asterisk to use the TDM400P card is called Zaptel, and if you worked your way through "Turn Your Linux Box into a PBX" [Hack #41], you've already got the Zaptel drivers installed. Since these drivers don't yet work with Mac OS X, you'll be able to use the TDM400P card only on i386 Linux. I'll cover enabling the drivers in a moment.

> If you're using certain servers, the TDM400P might not be compatible with certain Dell PowerEdge servers. Also, some Intel equipment has known issues with the TDM400P card. So check Digium's web site (*http://www.digium.com/*) to be sure: you could save yourself some time and aggravation. VoiceTronix makes alternative cards that you might want to consider, too.

But first, you need to get the card installed. This is pretty straightforward. If you've got a spare PCI slot, you're ready to snap the card into place. There are four numbered modules on the card, which correspond to the four numbered eight-wire jacks on the case plate of the card. Before inserting the card, screwing down the back brace, and replacing the PC's cover, you might want to note which jacks are for connecting phones and which are for connecting phone lines. It's hard to know once the case is on.

There's one more thing to connect after the TDM400P is slipped into place: a four-wire hard-drive power cable that runs from one of the PC's power leads to a power connector on the card. This cable brings power to the card above and beyond what's available from the PCI bus so that the card can provide ring voltage to any phones that are connected. You don't need to connect this little power cable if you aren't using FXS modules.

Once the card is in and your Linux box is booted up, you'll need to make sure the Zaptel drivers that were compiled when you first installed Asterisk are loaded before Asterisk is launched as a part of your normal system startup. To accomplish this, execute these commands before Asterisk is launched, perhaps in */etc/rc.d/rc.local*:

```
/sbin/modprobe zaptel
/sbin/modprobe wctdm
ztcfg
```

Note the difference between this startup routine, which provides driver support for both analog phones and analog phone lines, and that of "Connect a Legacy Phone Line Using Zaptel" **[Hack #44]**, which provides driver support only for phone lines. The addition of wctdm is the difference.

> Run make config in your Zaptel and Asterisk source directories to create startup scripts that are customized for your Linux distribution.

The *wcfxs* driver needs to be modprobed only if you're using FXO modules (with phone lines attached), and the *wctdm* driver needs to be modprobed only if you're using FXS modules (with analog phones attached). The *ztcfg* application tests the Zaptel driver configuration and returns nothing if it's valid. If there's a hardware error or a problem with the Zaptel configuration files, which we're about to discuss, *ztcfg* will return an error description to the standard error output.

The */etc/zaptel.conf* settings tell the Zaptel driver framework which modules on the TDM400P card are which. For a card with two FXO and two FXS modules, a configuration file like this would be used:

```
; zaptel.conf example
loadzone = us
defaultzone=us
fxsks=1,2
fxoks=3,4
```

The numbers assigned, 1 through 4, are channel numbers that Asterisk will use to refer to activity on each module. Within Asterisk, each legacy interface port on the TDM400P has its own channel. Since the point of this hack is to build a four-line phone server, we're going to assume that all of the channels are using the same type of signaling:

```
; zaptel.conf example
loadzone = us
defaultzone=us
fxsks=1,2,3,4
```

In this case, there are four FXO interfaces, to which we're going to connect one phone line apiece. fxsks (FXS Kewlstart) signaling is specified because the phone company switch to which these phones connect expects us to be using a telephone. Ordinarily, telephones interface to that phone-company switch (called a *foreign exchange office* in signal-ese) using electro-mechanical line signaling (called *FXS signaling* in Asterisk slang). Hence, our switch will be pretending to be a phone by using FXS signaling on its FXO interfaces. Confused yet? No worries—once your server is up and running and your TDM400P is functioning as planned, you'll probably never need to mess with it again.

Now that you've saved *zaptel.conf*, bring up */etc/asterisk/zapata.conf* in a text editor and make sure it resembles this example precisely:

```
[channels]
language=en
context=default
signalling=fxs_ks
usecallerid=yes
hidecallerid=no
```

```
callwaiting=yes
callwaitingcallerid=yes
threewaycalling=yes
transfer=yes
callreturn=yes
group=1
channel => 1
channel => 2
channel => 3
channel => 4
```

This block of config fine-tunes the settings of the four Zaptel channels that are provided by the card (for a concise description of what all of these settings do, check out Chapter 17 of O'Reilly's really amazing book, *Switching to VoIP*). Save this file before proceeding.

Set Up Incoming Calls

Now, pay a visit to */etc/asterisk/extensions.conf*. Here, you'll need to adjust the default context section of the file so that incoming calls on the four lines can be handled appropriately.

Take a look at this sample default context section in *extensions.conf*, which deals with incoming calls from the TDM400P-connected phone lines (and from any other channels that point to the default context):

```
[default]
exten => s,1,Dial(SIP/100,30)
exten => s,2,Voicemail(100)
exten => s,3,Hangup
```

So, as of right now (or at least after you reboot your Asterisk box or load the kernel modules manually), incoming phone calls to the connected phone lines will ring on the SIP phone configured as SIP peer 100. For a refresher on SIP peers, refer back to "Connect a Phone Line Using an FXO Gateway" [Hack #43].

Set Up Station-to-Station Calls

If you'd like your SIP phones to be able to call each other, be sure to add the following extension to the default context:

```
exten => _1XX,1,Dial(SIP/${EXTEN},30)
exten => _1XX,2,Voicemail(${EXTEN})
exten => _1XX,3,Hangup
```

The _1XX pattern matches any phone numbers dialed that are three digits long and begin with 1. It deals with them by attempting to ring the SIP peer that is registered with a user ID that matches the extension number dialed, and then sends them to the appropriate voicemail box after 30 seconds if the SIP peer doesn't answer.

Set Up Outgoing Calls

Now, you've got to make it so that any connected phones can place calls using the four lines that you've just hooked up to the installed FXO modules. This is accomplished by a special pattern-matching extension in those phones' contexts. For SIP phones, this is established in *sip.conf*. Let's say that a SIP phone's context is [private-phones]. To allow this SIP phone to dial out using your newly connected phone lines, you'll need to make sure there is a context in *extensions.conf* that looks something like this:

```
[private-phones]
exten => _NXXXXXX,1,Dial(Zap/g1/${EXTEN})
exten => _NXXXXXX,2,Congestion
exten => _1NXXNXXXXXX,1,Dial(Zap/g1/${EXTEN})
exten => _1NXXNXXXXXX,2,Congestion
exten => 911,1,Dial(Zap/g1/911)
exten => 911,2,Congestion
```

The string patterns _NXXXXXX and _1NXXNXXXXXX are actually masks designed to identify phone numbers that are 7 and 11 digits long, respectively. This way, if the dialed number is 7 or 11 digits long, Asterisk knows it must dial the number (represented by the variable ${EXTEN}) using the group of four phones (Zap/g1) you previously defined in *zapata.conf*. The 911 extension performs call routing to the phone company's Public Safety Answering Point (PSAP), via the standard 911 phone number. (In countries other than the United States, local jurisdictions will use different numbers for this purpose, so check with your local emergency dispatch authority to find out what number to use.)

Add a _011X extension to enable international dialing.

Of course, none of this is going to work until the drivers are loaded and the dial plan is reread by Asterisk, so give your machine a reboot, or load the modules and restart Asterisk manually. Then, call and be called—on the cheap. The coolest thing about the PBX you've just built is its cost effectiveness. To buy a four-line business phone system new is usually more expensive than equipping an Asterisk box like you've done in this hack. Plus, you've got access to Asterisk's programmable dial plan and application programming interfaces (the Asterisk Gateway Interface and Asterisk Manager API), giving you a metric ton more capabilities than a low-end commercial PBX.

Master Music-on-Hold

"Can you hold on a minute?" the operator asks. Suddenly you're listening to
Frank Sinatra singing "New York, New York." Before you know it, you're
tapping your finger and the wait doesn't seem so bad.

Few things are more dreaded among telephony end users than the short yet
foreboding phrase, "Please hold." Perhaps what bothers folks is that they
never know quite how long they're going to be on hold, or maybe it's the
notion that they're going to have to re-explain themselves to a whole new
person who winds up on the line after the hold time is over with. Fortunately, music-on-hold makes that wait time a little bit more tolerable.

Asterisk gets its music-on-hold sound signals from MP3 files that are
decoded and piped to Asterisk by one of two supported MP3 players:
Mpg123 and MPEG Audio Decoder (MAD). In this hack, I'm going to use
MAD because there are some well-documented security issues with Mpg123
that have yet to be dealt with. (Of course, if you'd like to use Mpg123, you
can just issue make mpg123 from your Asterisk source directory.) To get MAD,
start at *http://mad.sourceforge.net* for a list of mirror sites to download the
MAD distribution. You'll need three pieces: the ID3 library, the MAD
library, and the *madplay* application. Each is in a separate archive that you'll
need to download, unpack, and install as follows:

```
# cd /root
# mkdir mad
# cd mad
# wget http://kent.dl.sourceforge.net/sourceforge/mad/madplay-0.15.2b.tar.gz
# wget http://internap.dl.sourceforge.net/sourceforge/mad/libmad-0.15.1b.
tar.gz
# wget http://peterhost.dl.sourceforge.net/sourceforge/mad/libid3tag-0.15.
1b.tar.gz
# tar xvzf madplay-0.15.2b.tar.gz
# tar xvzf libmad-0.15.1b.tar.gz
# tar xvzf libid3tag-0.15.1b.tar.gz
# cd libid3*
# ./configure
# make
# make install
# cd ../libmad*
# ./configure
# make
# make install
# cd ../madplay*
# ./configure
# make
# make install
```

Now, type **madplay** and press Enter. If you get a "failed to load" message about one of the library files, as shown here, the installation routine might have put the libraries in the wrong location:

```
madplay: error while loading shared libraries: libid3tag.so.0: cannot open
shared object file: No such file or directory
```

If this is the case, try moving them. On my system, I had to move them to */usr/lib*:

```
# mv /usr/local/lib/libmad.so* /usr/lib
# mv /usr/local/lib/libid3tag.so* /usr/lib
```

Once *madplay* executes without any error notices, you're ready to go on to the next step. You've got to tell Asterisk's voicemail module that you want it to use *madplay* as its preferred player. Comment out the default line in the */etc/asterisk/musiconhold.conf* file, and add an entry like this in its place:

```
default => custom:/var/lib/asterisk/mohmp3/,/usr/local/bin/madplay \
--mono -R 8000 --output=raw
```

This tells Asterisk to use the *madplay* application to stream random MP3 files from */var/lib/asterisk/mohmp3* in mono at a forced sample playback rate of 8 MHz (perfect for telephony).

Though you don't need the Zaptel driver or card for a SIP-only setup, music-on-hold bridging is dependent on the Zaptel driver framework's built-in timing code, and you won't hear much music on hold unless you load either a real Zaptel driver (for a real Zaptel card) or the Zaptel *ztdummy* driver, which is meant to fill in on machines that don't have an actual Zaptel board installed. Lucky for you, when you compiled the Zaptel drivers [Hack #41], you also unwittingly compiled *ztdummy*. How convenient. Put these commands in */etc/rc.d/rc.local* before Asterisk loads if you have no Zaptel card installed:

```
modprobe zaptel
modprobe ztdummy
```

Next, make a test extension that lets you listen to some on-hold music. Place an entry like this in */etc/asterisk/extensions.conf* in the most appropriate context:

```
exten => 100,1,MusicOnHold(30)
exten => 100,2,Hangup
```

The idea here is that when you dial 100 in this context, you'll get 30 seconds of hold music before the server disconnects your call. Save the changes to */etc/extensions.conf*, and go ahead and reboot your Linux box (or modprobe *ztdummy* and restart Asterisk).

Hacking the Hack

You can assign different groups of phones and phone lines to their own music-on-hold classes (classes define selections of recordings that you can assign to groups of peers) so that they hear different music. A group of SIP phones can be in one music-on-hold class, and a group of Zaptel-connected phone lines can be in another.

Add as many classes as you like (such as default, as shown earlier) to the *musiconhold.conf* file, and then "point" your various Zaptel channels and SIP phones at those classes. For Zaptel channels, you'll configure this in *zapata.conf*. The first two Zaptel channels are pointed at the Stevie-Ray class, and the second two are in the BB-King class:

```
[channels]
language=en
context=default
signalling=fxs_ks
usecallerid=yes
hidecallerid=no
callwaiting=yes
callwaitingcallerid=yes
threewaycalling=yes
transfer=yes
callreturn=yes
group=1
musiconhold=Stevie-Ray
channel => 1-2
musiconhold=BB-King
channel => 3-4
```

Hacking the Hack Some More

If you'd like to use a streaming MP3 Internet radio station instead of a group of MP3 files for your on-hold music source, make an entry like this in *musiconhold.conf* to create a class:

```
default => /var/lib/streaming,http://64.236.34.196:80/stream/1040
```

Now, create the directory */var/lib/streaming* and leave it empty, and this class will play back the streamed audio after your next Asterisk restart.

HACK #51 Record Calls

Pitch the microcassette and stick-on microphone. With Asterisk, all you need to record a phone call is Monitor().

There are two ways to record calls with Asterisk. One way is to use a softphone that supports call recording or some other client-side desktop solution

(in fact, "Secretly Record VoIP Calls" [Hack #85] describes precisely this scenario). The other way is to have Asterisk do all the recording and have SoX
do all the mixing. SoX, short for SOund EXchange, is the Swiss army knife of
sound-conversion tools. It allows all kinds of format conversion, resampling,
and mixing, topics covered in more detail in "Create Telephony Sounds with
SoX" [Hack #24].

To record a call with Asterisk, you can use the built-in Monitor dial-plan
command. In *extensions.conf*, any extension can be monitored as follows:

```
exten => s,1,Answer
exten => s,2,Monitor(wav,most-recent-call,M)
```

This example creates a WAV file called *most-recent-call-ext* in */var/spool/
asterisk/monitor*. The M argument causes the call to be mixed automatically
so that caller and receiver can both be heard in the same file. Without the M,
Monitor would just create two different files, *most-recent-call-in-ext* and
most-recent-call-out-ext. *ext* represents the extension that the caller dialed to
trigger this Monitor to begin with.

> SoX must be installed for the M option to work. Without
> SoX, Asterisk cannot output automatically mixed call record-
> ings. Most of the major Linux distributions provide a SoX
> package as an installation option.

Hacking the Hack

If you want to keep every call you record without overwriting already-
recorded WAV files, you'll need to come up with an automatic way of
uniquely naming every file that Monitor creates. The best way to do this is
probably to base the filename off of the current system date and time. Not
only does this make them unique, but it also affords you an easy way to find
files by date and time later on when you need them. This example uses the
${DATETIME} variable to produce a file whose name is something like *112205-
09:45:42-40*:

```
exten => 40,1,Answer
exten => 40,2,Monitor(wav,${DATETIME},M)
```

Once the files are recorded, you can use cron to automatically archive them
with *gzip*, or even use the mail command to send them to an email address,
much as you did with faxes in "Build an Inbound Fax-to-Email Gateway"
[Hack #91].

Get Your Daily Weather Forecast from Your Telephone

The Weather Channel has "Local on the 8s" every 10 minutes, but why wait 10 minutes for your forecast when you can be listening to it on your IP phone right now?

Aside from cataloging sea species and running a really great tsunami readiness web site, the National Oceanic and Atmospheric Administration (NOAA) also operates the National Weather Service. Those are the guys from whom excitable TV meteorologists get their severe weather warning and watch information. But TV weather guys don't have an exclusive on NOAA's weather data feeds.

At NOAA's web site, *http://www.noaa.gov/*, localized weather data is published in text-file feeds that are updated regularly. Your Asterisk server can grab these feeds and, thanks to Festival **[Hack #92]**, read you a weather report based on their contents. Have a look at this example:

```
exten => 50,1,Answer
exten => 50,2,System(/usr/bin/curl -s \
ftp://weather.noaa.gov/data/forecasts/city/oh/cleveland.txt \
| text2wave weather-feed.wav
exten => 50,3,Wait(1)
exten => 50,4,Playback(/tmp/weather-feed.wav)
exten => 50,5,System(rm /tmp/weather-feed.wav -f)
exten => 50,6,Hangup
```

The extension 50 grabs the text feed for Cleveland, Ohio, using the *curl* application, and immediately converts it using *text2wave*, a piece of the Festival distribution, into a WAV, which it plays back using Asterisk. If you want to keep tabs on the weather in a few different cities, you can create an extension for each.

If you don't have *curl*, grab it from *http://curl.haxx.se/*.

Put a Happy Face on Asterisk Using AMP

When you've got Apache and MySQL on your Asterisk PBX, you've got the makings of a web-based administration interface for your whole phone system.

Since Asterisk runs on Unix, it is able to leverage many of the niceties of a modern Unix environment: shell scripts, Perl programs, sockets, and so on. Historically, one of the chief shortcomings of Unix—and of Linux in particular—is the lack of a graphical user interface (GUI). Asterisk shares Unix's general inferiority in the user-interface department. But there's something you can do about it.

Asterisk Management Portal (AMP) gives you some real interface power tools: a web-based configuration tool suitable for nontechnical administrators, database routines for storing and retrieving the PBX's dial plan, and some handy preconfigured call flow and fax features that make day-to-day life with Asterisk much easier. For instance, AMP lets you upload music-on-hold files using a web interface and lets you create IVR menus without having to type them directly into *extensions.conf* or to program Asterisk macros.

How AMP Works

AMP provides a web-based GUI using Apache and connects to Asterisk using a combination of techniques—most notably, via the Asterisk Manager API. It uses PHP to build the web pages you interact with, and it controls Asterisk with code written in Perl. MySQL provides a repository where the entire dial-plan configuration is stored, retrieved, and modified by the web interface.

The Setup Process

AMP has a ton of software prerequisites, as you can see. But it's fairly easy to install. The basic steps are spelled out here and are detailed in the following sections:

1. Get the prerequisites, including Apache and MySQL.
2. Install Perl modules and custom telecom tools.
3. Build the MySQL database.
4. Run AMP's install script and finish up.

Get the prerequisites. A few dependencies are standing between your Linux server and AMP. Check to make sure that your Linux box is running Apache, *libtiff*, MySQL with development libraries installed, PHP (version 3 or higher), OpenSSL, Perl, ncurses, SoX, and *curl*. If you're running a Red Hat 7 or later distribution, you should have all of these packages either pre-installed or available via RPM. If you're not using Red Hat, chances are still pretty good that you've got everything you need, because most of these packages are either commonplace or required by Asterisk, and therefore are already installed on your machine.

Before you can go any further, though, you need to be certain that your Asterisk instance is running as a nonroot user. To do so, follow the recommendations in "Run Asterisk Without Root, for Security's Sake" [Hack #54], because the rest of the AMP installation is going to assume your Asterisk instance runs as a nonprivileged user. But keep your finger on this page, because there's a lot more to do!

Install Perl modules and telecom tools. You can download AMP from *http://amportal.sourceforge.net*. Unpack the AMP source distribution using tar (there are numerous examples of tar-unpacking throughout this book) into the */usr/src/* directory. Once it's unpacked, install the Net: Telnet Perl module from CPAN, which allows Perl-based packages such as AMP to use Telnet sockets:

```
# perl -MCPAN -e "install Net::Telnet"
```

Now, to enable AMP's music-on-hold upload feature, you can use *vi* to make a few modifications to the PHP configuration (PHP 4 users might need to substitute */etc/php4/apache2/php.ini* in place of */etc/php.ini*). The idea here is to increase the upload_max_filesize value to 20M and to change the corresponding LimitRequestBody value to 20000000. This way, you'll be able to use AMP to upload large files, like music-on-hold MP3s.

```
# vi +482 /etc/php.ini upload_max_filesize=20M
# vi +14 /etc/httpd/conf.d/php.conf LimitRequestBody 20000000
```

If you've already modified your PHP configuration files, these commands will not work correctly, and you should find the appropriate lines manually instead.

Next, you'll need to install the Asterisk Perl modules for Asterisk, like this:

```
# wget http://asterisk.gnuinter.net/files/asterisk-perl-0.08.tar.gz
# tar xvgf asterisk-perl-0.08.tar.gz
# cd asterisk-perl-0.0.8
# perl Makefile.PL
# make all
# make install
```

Then, grab a couple more Perl modules from CPAN and install them (these enable the forwarding of faxes received, if you want AMP to handle faxes):

```
# perl -MCPAN -e "install IPC::Signal"
# perl -MCPAN -e "install Proc::WaitStat"
```

To make sure that AMP's email integration works correctly, grab a copy of the MIME Construct package from Roderick Schertler (*http://search.cpan.org/src/ROSCH/mime-construct-1.9*) and unpack it to */root* or */usr/src*, whichever you prefer. Then, from the directory where it's been unpacked, install it as follows:

```
# make Configure.PL
# make install
```

To add fax-receiving support to AMP, install the *spandsp* package per the instructions in "Turn Your Linux Box into a Fax Machine" [Hack #90]. Then, you'll need to set up the MySQL CDR interface for Asterisk. (When you downloaded the Asterisk CVS, this was downloaded to */usr/src/asterisk-addons*.)

```
# cd /usr/src/asterisk-addons
# make clean
# make
# make install
```

Configure the MySQL database. Now you're getting to the meat of the hack: the MySQL database for storing the CDRs and AMP's replica of the Asterisk configuration. To set this up, blow the dust off your latent MySQL skills, and issue the following commands:

```
# /usr/bin/mysql_install_db
# /etc/init.d/mysqld start (or /etc/init.d/mysql start)
# mysqladmin -u root password 'db_root_pwd'
# mysqladmin create asteriskcdrdb -p
# mysql --user=root --password=db_root_pwd asteriskcdrdb < \
/usr/src/AMP/SQL/cdr_mysql_table.sql
# mysqladmin create asterisk -p
# mysql --user root -p asterisk < /usr/src/AMP/SQL/newinstall.sql
```

The text files directed to MySQL's standard input are provided as a part of the AMP distribution, and they contain all the queries needed to set up AMP's database. Now, launch the MySQL client:

```
# mysql --user root -p
```

Once you get to the mysql prompt, you can begin entering the access privileges for the database:

```
mysql> GRANT ALL PRIVILEGES ON asteriskcdrdb.* \
TO asteriskuser@localhost IDENTIFIED BY 'amp109';

Query OK, 0 rows affected (0.00 sec)

mysql> GRANT ALL PRIVILEGES ON asterisk.* \
TO asteriskuser@localhost IDENTIFIED BY 'amp109';

Query OK, 0 rows affected (0.00 sec)

mysql> quit
```

Run AMP's install script and finish up. About the only conventional part of this configuration is AMP's shell script for installing its standard files, which you need to run now:

```
# /usr/src/AMP/install_amp
```

Finally, add */usr/local/lib* to the */etc/ls.so.conf* file, which will include the *spandsp* fax libraries you loaded earlier. Add the following lines to your */var/lib/asterisk/.bash_profile* file:

```
# PATH=$PATH:/usr/sbin:$HOME/bin
# export PATH
# export LD_LIBRARY_PATH=/usr/local/lib
```

Then open */etc/rc.d/rc.local* in your favorite text editor. Replace the line that currently loads Asterisk (probably something like asterisk -vvv &) with this:

```
/usr/sbin/amportal start
```

Are you still reading? Excellent! You've just installed AMP. Now, to try it out, you can restart Asterisk, Apache, and MySQL, or you can just reboot to achieve the same effect. Once your reset or reboot is done, point your browser to *http://AsteriskServerAddress*. You'll be greeted with the Asterisk Management Portal, the ultimate Asterisk GUI. Now, go have fun configuring. (If you want this page to be secured by a username and password, you can use Apache's *htpasswd* utility. For more info on this, check out *http://www.apache.org/*.)

Run Asterisk Without Root, for Security's Sake
HACK #54

Running a critical service as root makes a security-minded sysadmin squirm. But it doesn't have to be that way. Asterisk doesn't need to run as the all-powerful root user.

By default, Asterisk runs as root—the user account with total, unrestricted power. This is generally considered a bad idea, as an exploit to Asterisk can lead to someone taking over your entire machine. To avoid this, the Apache Web Server doesn't usually run as root. This hack shows you how to run Asterisk as a less-godly user.

To do so, create a user called *asterisk*. In the following command, I use the Red Hat adduser command:

```
# adduser -c "Asterisk PBX" -d /var/lib/asterisk asterisk
```

Next, you'll need to alter Asterisk's *Makefile*, located at */usr/src/asterisk/Makefile*. Using your favorite text editor, find the ASTVARRUNDIR constant in the file, and alter its definition to match what follows:

```
ASTVARRUNDIR=$(INSTALL_PREFIX)/var/run/asterisk
```

The directory referenced here needs to be writeable by the user running Asterisk, just as the directory normally used should be writeable only by root. By changing the setting, you're allowing Asterisk to use a directory that can be written by its own nonroot user account. Now, recompile Asterisk using this sequence of commands:

```
# cd /usr/src/asterisk
# make clean ; make install
```

Once the recompile and install are done, you'll need to make sure the new user account has appropriate permission to several Asterisk-related directories, including the one you referenced in the altered *Makefile*:

```
# chown -R asterisk:asterisk /var/lib/asterisk
# chown -R asterisk:asterisk /var/log/asterisk
# chown -R asterisk:asterisk /var/run/asterisk
# chown -R asterisk:asterisk /var/spool/asterisk
# chown -R asterisk:asterisk /dev/zap
# chmod -R u=rwX,g=rX,o= /var/lib/asterisk
# chmod -R u=rwX,g=rX,o= /var/log/asterisk
# chmod -R u=rwX,g=rX,o= /var/run/asterisk
# chmod -R u=rwX,g=rX,o= /var/spool/asterisk
# chmod -R u=rwX,g=rX,o= /dev/zap
# chown -R root:asterisk /etc/asterisk
# chmod -R u=rwX,g=rX,o= /etc/asterisk
```

You can now launch the Asterisk server from the new user account, or from root using the su command:

```
# su asterisk -c /usr/sbin/safe_asterisk
```

Finally, you'll need to adjust the *safe_asterisk* script so that it uses the new user account to launch Asterisk, rather than root. To do so, open */usr/sbin/ safe_asterisk* in your favorite text editor, and add **su asterisk -c** before each instance of an asterisk command. Be sure to leave the commands unchanged, aside from prefixing them with the su command.

Once these steps are taken, Asterisk will have only as much power as you grant the *asterisk* user. Would-be attackers might be able to crash Asterisk, but in so doing, they won't be able to gain access to root's credentials.

HACK #55 Link Two Asterisk Servers with PSTN

You don't have to have dedicated point-to-point lines to link two PBX systems. Just use the PSTN phone lines that are already connected to them, and you can simulate a direct link.

You can build a two-office unified dial plan using two Asterisk servers. This way, a user need only dial the extension of the user at the other office to reach him, instead of calling that office's main number, waiting for prompts, and then dialing the user's extension. Asterisk can handle all of these steps automatically, routing the call to the other office's PSTN trunk, waiting until it's answered, and dialing the recipient's extension to complete the connection. Figure 4-11 illustrates just such a configuration.

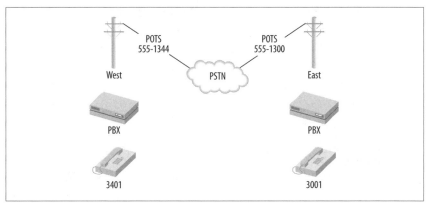

Figure 4-11. Two offices with PBXs connected to the PSTN

Ordinarily, if a West user wanted to reach an East user, he'd have to pick up his phone, dial the phone number of the East office, wait for an answer, and then request that user, either by speaking with a receptionist or by dialing that user's extension. This awkward process is shown in Figure 4-12. Direct Inward Dial could shorten the process, but the dialing user still wouldn't be able to reach his co-worker using a convenient, four-digit extension.

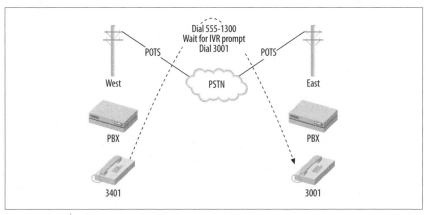

Figure 4-12. A caller has to dial a lot of digits to reach his intended recipient at the other office

The Configuration

We'll use the same dial-plan extension-numbering convention as shown in
Figure 4-12. Phones at the East office will be 3000–3099; phones at the
West office will be 3400–3499. Have one SIP phone register with the East
server and the other SIP phone with the West server (our West user will be
using the phone that registers with the West server, and the East user will be
using the East-registered phone). The following code shows the SIP peer
config for 3001 at the East office:

```
# East office sip.conf
...
[3001]
callerid="East User" <3001>
canreinvite=no
context=default
host=dynamic
mailbox=3001
secret=3001
type=friend
username=3001
```

The following code represents the SIP peer config for 3401 at the West
office:

```
# West office sip.conf
...
[3401]
callerid="West User" <3401>
canreinvite=no
context=default
host=dynamic
mailbox=3401
secret=3401
type=friend
username=3401
```

With these first two configs committed, the SIP phones can now register
with their respective Asterisk servers and place calls in their own default
contexts. But they still can't call each other without dialing a lengthy PSTN
phone number, waiting for the auto-attendant, and dialing the extension on
the answering Asterisk system. To get around that, we can tell both Asterisk
servers to route calls bound for the extension-number range of the other
office out through the PSTN and automatically dial the extension on the
answering system, as follows. We'll start with the dial-plan config for the
East office:

```
# East office extensions.conf
...
[default]
exten => _34XX,1,Dial(Zap/1/5551340,35,mD(${EXTEN}))
```

And we'll institute a mirror of that config so that West office users can dial 30XX extensions:

```
# West office extensions.conf
...
[default]
exten => _30XX,1,Dial(Zap/1/5551300,35,mD(${EXTEN}))
```

Let's dissect this exten directive. First, _30XX is a wildcard expression that matches any number dialed that begins with 30. 1 is the extension priority. The Dial command tells Asterisk to dial the number of the other office on the Zap/1 channel and to wait for up to 35 seconds for the call to be answered. Then, the D(${EXTEN}) option tells the Dial command to send DTMF digits representing the extension number that was dialed by the user. ${EXTEN} is an Asterisk variable that always contains the extension number used for the current call. Finally, as with all Dial commands, the call will be connected after the DTMF digits are sent.

The net result of this config is that the users at West can dial 3001–3099 to reach the users at East, and the users at East can dial 3401–3499 to reach the users at West, all without any PSTN dialing or auto-attendant interaction. Here, the PSTN trunks are used like private trunks to connect two switches, while the dial plan makes it easy for the users.

Control Caller ID When Using PSTN Trunks

In the preceding example, the receiving PBX doesn't know the extension number of the party who is calling, because the calling PBX supplies the caller ID signals for the Zaptel channel and phone line being used, *not* the caller ID signals for the extension that originated the call. So, the receiving user will see that she is getting a call from the "other office," but she won't know which user is calling her.

Using some Asterisk dial-plan wizardry, you can preserve the original caller's caller ID information throughout the interswitch calling process:

```
# West office extensions.conf
...
[default]
exten => _30XX,1,SetCIDNum(${EXTEN})
exten => _30XX,2,Dial(Zap/1/5551300,35,mD(${EXTEN}))
```

In this case, Asterisk will supply the originating extension number as the caller ID number. SetCIDNum establishes the caller ID number for outgoing channels on the current extension. This config would, of course, have to be mirrored for the East office, too:

```
# East office extensions.conf
...
```

```
[default]
exten => _34XX,1,SetCIDNum(${EXTEN})
exten => _34XX,2,Dial(Zap/1/5551340,35,mD(${EXTEN}))
```

To override the phone company's caller ID, you'll need to be using PRI signaling, and the phone company will have to permit you to supply your own caller ID information.

There you have it: a two-office Asterisk PBX network that uses existing telephone lines to simulate a direct link between the two sites.

Link Several PBXs over the Internet

Since wide area networking is the cornerstone of IP networking, VoIP can be extended outside the local area network. Six PBXs on six continents all managed by one person? No problem.

One of the beauties of VoIP is the last two letters—IP. IP stands for Internet Protocol, and IP is the protocol that makes the Internet and private wide area networks (WANs) possible. It provides the fundamental addressing and routing scheme that keeps data traffic flowing around the globe.

Just as people have been setting up VPNs to link remote offices' LANs over the Internet, you can now use the Internet to link several remote PBXs to create one large interconnected voice network. This will enable extension-to-extension calling from one office to another, over the Internet, all at no cost per call. Asterisk is the perfect solution for this purpose.

Keeping VoIP secure as it travels the globe is no simple matter. You can use Internet-based VPNs, but they can degrade quality. A more expensive and more reliable alternative to secure global VoIP trunks might be a managed VPN service or a frame relay service.

I will assume that you already have Asterisk in use as the PBX at all of your offices. If this is not the case, you might want to look into setting up an Asterisk gateway machine—a server that provides VoIP enablement for a legacy PBX (see "Build a Four-Line Phone Server" [Hack #49] and "Connect a Legacy Phone Line Using Zaptel" [Hack #44] for ideas on how to use legacy TDM interface cards).

Now let's assume:

- You have three offices: one in Chicago, one in Tokyo, and one in London.
- None of the offices has more than 99 separate extensions.

- They all have 24/7 Internet connectivity.
- They all have static IP addresses.
- Their Asterisk installs are directly on the Internet, or the network administrator has forwarded/passed UDP port 4569 to the Asterisk server in each location.

It is worth pointing out that the last three items do not necessarily have to be the case. With all of Linux's power, you can actually work around those issues. While this is too much information for me to cover here, you can use the "register" feature of IAX with dynamic IP addresses, even those behind NAT! A simple Google search should return the necessary details to use register statements successfully to work around those problems.

Let's also assume that none of your locations has overlapping extensions. That is, each location has globally unique extension numbers. For the purposes of this hack, we are going to assume that your extensions are set up like so:

Chicago
> 81XX (where XX is 01–99)

Tokyo
> 82XX

London
> 83XX

So your first extension in Chicago is 8101, and your last possible extension in London is 8399.

If your extensions are not set up like this, it will probably be to your advantage to renumber. As you will see, this method has one big advantage: unless every extension is globally unique, it's more difficult for each Asterisk server to route calls.

The beginning "8" signifies an internal extension. I begin with this to standardize on four-digit extensions so that an internal extension is readily recognized as being a free internal call, as opposed to an outside call to the pizza place. (This prefix helps Asterisk figure out where to route the call—to an internal user or to the phone company.) Using the same numbering convention around the world will make your life easier. When you bring that new office in Stockholm online, you just have to assign it the 84XX range and update your Asterisk servers, and the phones around the world will automatically recognize it as a valid range.

If you have not already done so, let's set up some basic DNS records for this system. We are going to create several A records in our existing DNS zone, *twidgets.com*. These A records are going to be called *chicago.twidgets.com*, *tokyo.twidgets.com*, and *london.twidgets.com*. They should each point to the static IP address of the Asterisk server at each respective location.

Once DNS is set up properly, verify basic IP connectivity by using the ping command to each location, from each location. Ping Tokyo from Chicago and London. Ping London from Tokyo. You get the drift. This is what you should have so far:

City	Hostname	Extension block
Chicago	*chicago.twidgets.com*	81XX
Tokyo	*tokyo.twidgets.com*	82XX
London	*london.twidgets.com*	83XX

Configuring the Dial Plan

On each Asterisk server, you need to add a matching extension for each dial pattern. So, log into your server in Chicago and add the following to your [internal] context in */etc/asterisk/extensions.conf*:

```
exten => _82XX,1,Dial(IAX2/guest@tokyo.twidgets.com/${EXTEN},20)
exten => _82XX,2,Congestion
exten => _83XX,1,Dial(IAX2/guest@london.twidgets.com/${EXTEN},20)
exten => _83XX,2,Congestion
```

Let's take a look at what we have done so far. In the first line, we're telling Asterisk to create an extension that matches anything in the 8200–8299 range. (Remember those *X*'s from before? They signify to Asterisk any digit between 0 and 9.) The first thing that Asterisk should do is try to reach that extension at the Tokyo office by using the IAX protocol (Version 2) with the username *guest*. The *guest* username is just a placeholder. You can use a more descriptive name if you want.

> The IAX protocol is a signaling protocol, like SIP, which is more efficient at trunking multiple simultaneous calls between the same two locations.

If that extension at the Tokyo office is unreachable for any reason, Asterisk will return congestion. Congestion is usually signaled to the user as what is called "fast-busy." If you have ever left a POTS phone off the hook for too long, you have heard a fast-busy.

Adding the Remote Locations

Save *extensions.conf* and reload Asterisk with **asterisk -rx reload**. If the servers in Tokyo and London have been set up with those extensions, go ahead and try calling them. They won't be able to call you back yet, but you should at least be able to verify that you now have direct dial around the world (for free)!

You should now repeat this process on your Asterisk servers in London and Tokyo. For brevity's sake, I will give abridged versions of the preceding instructions for the Tokyo and London offices.

For Tokyo, edit */etc/asterisk/extensions.conf* and add the following to your internal context:

```
exten => _81XX,1,Dial(IAX2/guest@chicago.twidgets.com/${EXTEN},20)
exten => _81XX,2,Congestion
exten => _83XX,1,Dial(IAX2/guest@london.twidgets.com/${EXTEN},20)
exten => _83XX,2,Congestion
```

Save *extensions.conf* and reload Asterisk with **asterisk -rx reload**.

For London, open */etc/asterisk/extensions.conf* and add the following to your internal context:

```
exten => _81XX,1,Dial(IAX2/guest@chicago.twidgets.com/${EXTEN},20)
exten => _81XX,2,Congestion
exten => _82XX,1,Dial(IAX2/guest@tokyo.twidgets.com/${EXTEN},20)
exten => _82XX,2,Congestion
```

Save *extensions.conf* and reload Asterisk with **asterisk -rx reload**.

Hopefully, after this hack, you have realized that with four lines of configuration on three Linux boxes around the world, Asterisk can revolutionize the way your organization communicates. What would have been incredibly difficult and expensive to do just a few years ago has now been reduced to a few pages in a book. It's truly amazing!

—*Kristian Kielhofner*

Route Calls Using Distinctive Ring
#57
Do you have only one phone line, but wish you could use two phone numbers with your Asterisk server? Try distinctive ring.

Distinctive ring is a feature offered by some phone companies that permits you to use two or three phone numbers with the same POTS line. Depending on which number is dialed, the ring signal will differ, causing the ring to sound unique for each number. This feature allows parents to avoid answering their teenagers' incoming calls. With a fax/voice ring switch device, you can use distinctive ring as an inexpensive way to receive both fax and voice calls on a single line.

Distinctive ring is a legacy signaling solution. That is, it works only with POTS. On VoIP trunks, such functionality would be handled by out-of-band signaling.

With Asterisk, you can use distinctive ring to route calls automatically from the PSTN trunk to a specific phone or group of phones. Or, the distinctive ring can just be passed through to all of the phones on the private network, which will ring distinctively, and the intended recipient can answer her call on any available phone.

You can configure each Zaptel channel to detect up to four different distinctive signals. The first thing you'll need to do is open *zapata.conf* and add this configuration to the section for the trunk in question:

```
usedistinctiveringdetection=yes
```

Enabling distinctive ring on a Zaptel channel will cause a slight delay before Asterisk can answer incoming calls, because the distinctive ring signals can take up to five seconds for the Zaptel channel to detect.

The signals used by distinctive ring consist of analog electrical cadences—variations in voltage that cause analog phones to produce certain ring patterns. Asterisk uses the `dring` attribute in *zapata.conf* to describe the signals. Unfortunately, these signals vary from one regulatory jurisdiction to the next, and you'll have to figure out what value to give `dring` attributes yourself.

Here's how. When an incoming call is received on a POTS interface, Asterisk records the ring pattern in Asterisk's verbose logging output (assuming you launched Asterisk with -vvvv on the command line). Use the `tail` command with its -f option to watch your logfile for changes as they occur:

```
# tail -f /var/log/asterisk/full
```

While `tail` is following the logfile, call each number that causes distinctive rings on your POTS lines. When the POTS interface senses the ring pattern, a log entry will appear containing Asterisk's representation of it: a string of digits made up of three values separated by commas. Each value represents a duration of ringing, such that each ring pattern could have up to three rings of varying length in a one- or two-second time span. The pattern repeats at regular intervals until the call is answered.

This string supplies a value to the `dring` argument in *zapata.conf*. Repeat this process until you've identified the strings needed for each phone number associated with your POTS line. Here's a sample config in *zapata.conf* that describes two distinctive ring signals and assigns them different contexts in the dial plan:

```
usedistinctiveringdetection=yes
dring1=325,95,0
dring2=95,0,0
dringcontext1=TedsCalls
dringcontext2=JakesCalls
channel =>1
```

 Distinctive ring features outside North America can use caller ID signaling instead of ring-pattern signaling to indicate which phone number is being called. Check with your telephone company to see how they support distinctive ring.

The Zaptel channel's configuration will tell Asterisk the context into which distinctively rung calls are sent. In this example, we've used a POTS line with two ring signals and two corresponding contexts. Now, we've got to create those contexts in the dial plan. Here's a sample that accomplishes that in *extensions.conf*:

```
[TedsCalls]
exten => s,1,Dial(SIP/201,30)
exten => s,2,Voicemail(201)

[JakesCalls]
exten => s,1,Dial(SIP/202,30)
exten => s,2,Voicemail(202)
```

There! Ted's distinctive ring will send Ted's calls to SIP/201, and Jake's distinctive ring will send them to SIP/202.

 ## Tune Up Your Asterisk Logs

#58 How much log detail is too much? That depends on whom you ask. Asterisk's log output can be pretty granular, which is bad for disk utilization and good for troubleshooting.

Log analysis should be the core of your daily system monitoring and security activities. Like other softPBX servers, Asterisk supports flexible logging, providing several levels of logging detail in several different files. It also supports using syslog.

 By default, Asterisk stores its logs in */var/log/asterisk*.

You configure Asterisk logging in the */etc/asterisk/logger.conf* file, which Asterisk reads at boot time or whenever it is started. The first section of the file is [general], where you can assign a value to the dateformat option to specify what date format to use in Asterisk's logs. To figure out the syntax of the data formats, read the manpage for strftime( ) by running man strftime.

The next section, [logfiles], describes which files should be used for logging output and how detailed each should be. The syntax for this section is:

```
filename => level,level,level...
```

Consider the following logging configuration:

```
[general]
[logfiles]
messages.log => notice,warning,error
debug.log => notice,warning,error,debug,verbose
```

In this example, *messages.log* will contain a digest version of Asterisk's logging output, and *debug.log* will get everything in minute detail. Be careful with logs, though—Asterisk won't start once the logfiles reach 2 GB in size. On a busy system, a file like the preceding *debug.log* would hit that size pretty quickly, so make sure your logfile rotation includes Asterisk.

If you use console as a logfile name, Asterisk will assume you mean the console device, not an actual logfile. So, if you add this to the [logfiles] section, the desired level of logging will be output to the console session where Asterisk is launched:

```
[logfiles]
console => warning,error
```

Some attackers cover their tracks by removing commonly used logfiles that could contain evidence of their tampering with the system. So it's generally a good idea to keep logfiles in a nondefault place. That way, if an attacker uses an automated program to remove logfiles, the program will be less likely to find and destroy Asterisk.

If you were an intruder and wanted to control Asterisk, you might start by attacking the Asterisk Manager—the remote API that allows users who've provided the right password to control certain aspects of Asterisk's operations via a TCP connection. If your Asterisk server is open to the Internet, pay special attention to Asterisk Manager log entries.

To change Asterisk's default log location, edit */etc/asterisk/asterisk.conf* and change the astlogdir directive to a path of your choosing. (Then make sure that path has appropriate permissions to allow Asterisk to write files in whichever path you choose.) A sample *asterisk.conf* follows:

```
[directories]
astetcdir => /etc/asterisk
astmoddir => /usr/lib/asterisk/modules
astvarlibdir => /var/lib/asterisk
astlogdir => /var/log/asterisk
astagidir => /var/lib/asterisk/agi-bin
astspooldir => /var/spool/asterisk
astrundir => /var/run/asterisk
```

Syslog can be a target for Asterisk logging output, too. To enable it, use a syslog keyword in the [logfiles] section, similar to the console keyword:

```
syslog.local0 => warning,error
```

C H A P T E R   F I V E

Telephony Hardware Hacks
Hacks 59–71

One of the reasons VoIP is such a positive evolutionary step for telecommunications is that it employs a highly distributed, software-centric design philosophy. It has the extensibility and programmability of the Internet, putting telephony power back into the hands of the users, not the phone companies. It is programmability—*software*—that makes IP telephony such a killer application.

Yet, critical parts of voice telecommunications are entirely in the domain of *hardware*. This chapter focuses mainly on hardware hacks: projects with a decidedly piquant, earthy flavor; projects that deal with analog telephone adapters (ATAs), phone line gateways, and "bat phones." You won't need much in the way of 'l33t coding skills, but you'll use some basic Perl. Have some Ethernet patch cables and Velcro handy. Oh, and it might help to have a bottle of XML on ice if the mood is right.

In this chapter, you're going to be looking at VoIP and legacy telephony hardware—everything from state-of-the-art Internet Protocol (IP) phones to vintage rotary-dial candlestick phones—and how to get them working together. You might also pick up some handy tips for your cell phone, as well.

Record Calls the Old-Fashioned Way
#59
Digital and IP phone handsets are analog inside, which means you can use a transducer microphone to record a phone call.

It's fairly easy to record from a standard telephone using an inline recorder switch. These devices allow you to record the analog audio signal on a standard Plain Old Telephone Service (POTS) line or a handset-to-deskset line using an analog recording device like a microcassette recorder or a personal dictation recorder. An easy way to use one of these recorder switches to produce a digital recording of a call is to connect the mono audio output to your computer's microphone line-in.

Most of these switches (such as Radio Shack's model 43-1237) offer a 1/8-inch male audio connector, which is perfect for use with a PC sound card or a Mac line-in, which both tend to be 1/8-inch female connectors. Just plug the recorder switch into the phone line and your computer's audio input, and you've got an instant call recorder in the form of your favorite audio recording program (such as Windows Sound Recorder).

Since inline recorder switches work only with analog lines, you can't use them to record calls on digital or IP telephones. If you want to use your PC to record from these devices, you'll need something a little more James Bond-ish, like a transducer pickup. This is a microphone that you stick to the outside of your telephone handset, on the back of the receiver, that is sensitive enough to record the audio inside the handset. Since even digital and IP telephones use purely analog handsets, a transducer pickup can record them all. Some pickups (such as Radio Shack's model 44-533) include a built-in suction cup that adheres easily to the handset. Like a recorder switch, these pickups provide a 1/8-inch mini plug that you can mate with your sound card's audio line-in jack to make digital recordings.

 Recording phone calls can get you in trouble unless all parties on the call are aware that the call is being recorded. Check your local laws before recording any phone calls.

See Also

- "Record VoIP Calls on Your Windows PC" [Hack #13]
- "Record an Audio Chat on Your Mac" [Hack #23]

HACK #60 Make IP-to-IP Phone Calls with a Grandstream BudgeTone

With minimal effort, Grandstream's BudgeTone series of IP phones can make and receive calls on your network—even without a PBX server.

In most enterprise VoIP setups, you have a PBX that connects all of the phones on the network. The PBX acts as a centralized signaling authority and access-control server for all of the telephone users. But some IP phones don't need a PBX at all. They can call each other directly by way of an IP address. You're about to make a direct IP-to-IP call with a BudgeTone hardphone.

> A hardphone is an IP phone that isn't a softphone. It looks
> just like an ordinary business phone, but plugs directly into
> an Ethernet local area network (LAN).

The Grandstream BudgeTone 100 phone model has a Menu key, an LCD
display, and two arrow keys that you use to navigate its configuration menu
options: DHCP, IP Address, Subnet Mask, Router Address, DNS Server
Address, TFTP Server Address, Codec Selection Order, SIP Server Address,
and Firmware Versions (called *Code Rel* on the phone's screen). When you
get to the option you want, you press the Menu key to select it, and then
enter the numeric data required for each option using the keypad. Use this
menu only to set up the IP address, subnet mask, and router (default gateway) address, because you'll be doing the rest of the phone's configuration
using its web interface.

To get the phone enabled for the next configuration step, turn DHCP off
and assign an IP address, subnet mask, and router address.

You can perform more advanced configuration using the BudgeTone's built-
in web configuration tool. When you use your web browser to access the IP
address you assigned, you'll be prompted to log in to the phone. The default
password is *admin*.

Then, you'll be confronted with a big page of configuration options, many
of which are available only through this interface, not from the phone's keypad menu. For this project, the only settings we're concerned with are the
codec selection ones. Configure the first (highest-priority) codec to be
"PCMU" if you're in North America or "PCMA" if you're elsewhere in the
world. That's all we're going to cover about codecs for now. After you apply
any configuration changes, you need to power-cycle the BudgeTone.

Some IP phones offer a Telnet interface rather than a web-based one. To use
these tools, you must connect to the phone with a Telnet client rather than a
web browser. In any event, once the network configuration is set on the IP
phone, ping its address from another host on the same network subnet to
make sure it's speaking Transmission Control Protocol/Internet Protocol
(TCP/IP).

> Many VoIP devices need access to a time clock. The network
> time protocol (NTP) server we've chosen is *time.nist.gov*.
> More NTP servers are available from the list at *http://www.*
> *nist.gov/*.

Make an IP-to-IP Phone Call

With both IP phones connected to the same Ethernet switch, or directly connected (to each other) using a crossover patch cable, make a note of the IP address you've established for each. In this example, we'll use 10.1.1.103 for the receiver and 10.1.1.104 for the caller. If you have your phones configured for DHCP, give them this static configuration instead.

The BudgeTone can place IP telephone calls from one IP endpoint directly to another without the need for a VoIP call-management server. This is known as *IP-to-IP calling*. Since each IP phone has a unique identification characteristic within the scope of the network—an IP address—one phone can call the other by IP address as if it were a phone number.

Now, to dial by IP address. All IP addresses are 12 decimal digits long, even if preceding zeros aren't visible when notated. Conversely, the dots (.) that are normally included in a notated IP address are *not* dialed. So, on the BudgeTone phones, 10.1.1.103 is dialed as:

```
010 001 001 103
```

To dial, take the phone off the hook so that you hear a dial tone, and then press the Menu key, dial the address of your second phone according to the convention just shown, and press the Send or Redial button. Of course, nobody would want to dial 12-digit IP addresses to place phone calls all the time; call management servers, like Session Initiation Protocol (SIP) registrars, provide more elegant dialing conventions. However, dialing by IP address does allow you to circumvent call management and make a direct VoIP connection between two endpoints.

When the receiving phone rings, have somebody answer the call. If you can hear them talk through your IP phone's handset, you've just made your first successful VoIP phone call—sort of the IP equivalent of Bell and Watson's first phone call back in 1876.

If the receiving phone doesn't ring, you might have to check the IP address you dialed, check the phone's configuration to make sure it is listening on the default port for SIP—5060—and make sure SIP registration is turned off. These options concern the Grandstream's use with a PBX server, which isn't a factor in this case.

Mounting the Grandstream on the Wall

For practical, day-to-day use, Grandstream, shown in Figure 5-1, has a few shortcomings. At the top of my "bug list" for the Grandstream 101 is its half-baked support for being hung on the wall (in addition to sitting flat on the desktop). I say *half-baked* because Grandstream provides screw holes for hanging the phone on the wall, but it doesn't provide a notch to keep the handset on the phone when it's hanging. So, the handset just slips off the phone when you attempt to set it upright.

Figure 5-1. The Grandstream BudgeTone is a great, cheap SIP phone

This won't do. I've envisioned two ways to deal with this problem. First, you can go the Velcro route. Apply about a square centimeter of Velcro adhesive hook strip to the handset, at the point where a normal wall-hanging handset's notch would be. At the corresponding position on the phone itself, put the same amount of Velcro latch strip so that when you hang up the handset, it actually stays in place.

The second way to deal with the wall-hanging problem, which is probably a longer-lasting or more durable approach, is to drill a small hole in the phone base at the point where the wall-hanging notch should be. The hole should be about one-third of an inch to three-quarters of an inch in diameter. Then, again, at the corresponding spot on the handset, screw into the plastic casing a very short, round-headed screw. The head of the screw, if small enough to fit into the hole you drilled, should keep the handset firmly latched onto the phone's base. Not pretty, but it works.

HACK #61 Build a Custom Ringtone for Your Grandstream Phone

Sure, your cell phone has a custom ringtone, but does your IP phone? With a little help from Perl, you'll able to load any sound you like onto the Grandstream phone.

If you carry a cell phone, you've no doubt changed your ringtone once or twice. From a sample of a vintage mechanical ringer to a recording of a C-3PO line from *Star Wars*, ringtones have become central to pop-culture communication. So why can't you customize the ringtone on a Grandstream IP phone, one of the cheapest and most popular SIP hardphones available?

Well, since you asked, you *can*. It just takes a little hack job.

The Grandstream's firmware stores the ringtone in its own odd format, a uLaw sound file with a custom header at the beginning of it. It's simple enough to make a uLaw sound file; just use SoX [Hack #24]. But to add the header, a little Perl magic is needed.

The Code

This script was written by Tony Mountifield, and its purpose is to create a Grandstream-compatible ringtone file:

```
#!/usr/bin/perl
$filename = shift or die "need output filename\n";
undef $/;    # slurp whole file at once...
$audio = <>;    # ... like this
$filesize = 512 + length $audio;
if ($filesize & 1) {
    # length odd, add a zero byte (should never happen)
    $audio .= chr(0);
}
die "Audio file too large\n" if $filesize > 65536;

# this is the format for the header
$headerfmt = "n n n C4 n C C C C a22 n x216 n n x36 a216";
```

```
# get the current date and time
($min, $hour, $day, $month, $year) = (localtime)[1..5];
$year += 1900;
$month += 1;

# create the header, with zero for the checksum
$header = pack $headerfmt,
        0,          # 0000
        $filesize/2,
        0,          # put checksum in later
        1,0,0,1,    # version
        $year, $month, $day, $hour, $min,
        $filename,
        0,          # 0000 or 00C8 - why?
        256,        # 0100
        $filesize/2,
        "Grandstream standard music ring";

# sanity check
$headerlen = length $header;
die "header length wrong ($headerlen)\n" unless $headerlen == 512;

# add the audio
$header .= $audio;

# compute the checksum
$checksum = unpack "%16n*", $header;
#printf "checksum before = %04x\n", $checksum;

# insert it in the correct place
substr($header,4,2) = pack "n",-$checksum;

# ensure the new checksum is zero
$checksum = unpack "%16n*", $header;
#printf "checksum after  = %04x\n", $checksum;
die "checksum failed\n" unless $checksum == 0;

# write the file
open F, ">$filename" or die "can't open output file $filename: $!\n";
print F $header;
close F;
```

Running the Code

To use this program, save it as *makering.pl*, make it executable (chmod 755
makering.pl), and pipe a uLaw sound file into it in a shell, like so:

```
$ sox my_sound -r 8000 -c 1 -t ul - rate | makering.pl ring1.bin
```

In this example, the file *my_sound* will be resampled to 8000 Hz and will be
piped in uLaw format to the standard input of *makering.pl*, which is the Perl
script shown earlier. The enhanced output is then saved as *ring1.bin*. Upload

this file to the */tftpboot* directory of your Grandstream's TFTP server and then reboot your Grandstream. (For some tips on setting up a TFTP server, see "Make IP Phone Configuration a Trivial Matter" [Hack #80].) With a fun new ringtone, your IP phone is now as cool as your cell phone.

Tweak Your Sipura ATA
HACK #62

If you own a Sipura ATA, you've got a veritable softPBX hiding in that slick plastic enclosure. If only you knew how to set it up!

Sipura Technology, now a division of Cisco, makes some very powerful telephony devices. With hundreds of options and many potential combinations, literally thousands of possible configurations are available. While I can't cover them all here (for obvious reasons), I can give you a few examples to get your mind working.

Configure the Sipura by Dialing

Sipura's line of products has a powerful interactive voice-response (IVR) system built in that gives you administrative access to many of the ATA's features. In fact, the IVR will probably be one of your first experiences with Sipura's ATAs. The IVR (like the web interface) has quite a few options. Thankfully, Sipura publishes a user guide that details all of the available options in the IVR menu, as well as in the web configuration screens. In fact, so many options are available that the user guide was 87 pages long at the time of this writing! You can find this document in the Support section of *http://www.sipura.com/*.

After unpacking the Sipura and connecting the cables, you should pick up your phone and dial ****. This will connect you to the Sipura Configuration Menu. You will be asked to enter an option. But what option to enter? Table 5-1 will get you started.

Table 5-1. Configuration options for the Sipura, via IVR

Option name	Option number	Valid options	Notes
DHCP status	100	None	
Check IP address	110	None	Reads current IP
Set static IP address	111	IP address; enter IP, using * to input periods	
Check network mask	120	None	
Set network mask	121	Same as Set static IP address	

Table 5-1. Configuration options for the Sipura, via IVR (continued)

Option name	Option number	Valid options	Notes
Check gateway IP	130	None	Reads current IP
Set gateway IP address	131	Same as Set static IP address	
Check DNS server	160	None	
Set DNS server IP address	161	Same as Set static IP address	
"User" reset	877778	None	Resets all of the "user-change-able" settings to their defaults. Use with caution!
"Factory" reset	73738	None	Resets all of the available configuration options to their defaults. Use with caution!

Remember to add a trailing # for each option. So, for example, to have the Sipura read its current IP address, you should enter **** and then **110#**. Another thing to remember: when you are entering IP addresses for the device, default gateway, and DNS server, use the * key to represent periods (.). So, you'd enter the IP address 10.1.1.50 as 10*1*1*50#.

Various Tweaks

After you have the Sipura connected to your network and you know its IP address, you can get to its web interface. If you're used to dealing with web interfaces for Network Address Translation (NAT) firewalls and the like, the Sipura web interface, shown in Figure 5-2, is probably unlike anything you've seen. Most people that use the web interfaces of small-office/home-office (SOHO) NAT/firewall/router devices are shocked when they see the web interface on a Sipura. Here, I will attempt to point out the most common and useful, yet often overlooked, web interface parameters.

To reach the web interface, simply enter the Sipura's IP address in your web browser. Once you see the gray status screen, click the Admin link in the top right-hand corner. When the page refreshes, click Advanced. You should see several more tabs appear. Now we are ready!

While I fully encourage you to review the user guide and browse the configuration pages, I have summarized in Table 5-2 my "Top 10 Sipura options" for your hacking pleasure.

SIPURA
technology, inc.

Sipura Phone Adapter Configuration

| Info | System | User 1 | PSTN User | | Admin Login   basic | advanced |

System Information

DHCP:	Enabled	Current IP:	192.168.0.81
Host Name:	SipuraSPA	Domain:	kielhof.com
Current Netmask:	255.255.255.0	Current Gateway:	192.168.0.1
Primary DNS:	192.168.0.1		
Secondary DNS:			

Product Information

Product Name:	SPA-3000	Serial Number:	88012DA06633
Software Version:	2.0.11(GWg)	Hardware Version:	2.0.1(5886)
MAC Address:	000E08CAE53A	Client Certificate:	Installed

System Status

Current Time:	6/3/2005 02:45:17	Elapsed Time:	1 day and 08:20:07
Broadcast Pkts Sent:	0	Broadcast Bytes Sent:	0
Broadcast Pkts Recv:	5169	Broadcast Bytes Recv:	473913
Broadcast Pkts Dropped:	0	Broadcast Bytes Dropped:	0
RTP Packets Sent:	20218	RTP Bytes Sent:	4794432
RTP Packets Recv:	30047	RTP Bytes Recv:	4770680
SIP Messages Sent:	17628	SIP Bytes Sent:	6296556
SIP Messages Recv:	17761	SIP Bytes Recv:	6485410
External IP:			

Line 1 Status

Hook State:	On	Registration State:	Not Registered
Last Registration At:		Next Registration In:	
Message Waiting:	No	Call Back Active:	No
Last Called Number:	ata1@192.168.0.132:5060	Last Caller Number:	ata1
Mapped SIP Port:			
Call 1 State:	Idle	Call 2 State:	Idle
Call 1 Tone:	None	Call 2 Tone:	None
Call 1 Encoder:		Call 2 Encoder:	
Call 1 Decoder:		Call 2 Decoder:	
Call 1 FAX:		Call 2 FAX:	
Call 1 Type:		Call 2 Type:	
Call 1 Remote Hold:		Call 2 Remote Hold:	
Call 1 Callback:		Call 2 Callback:	
Call 1 Peer Name:		Call 2 Peer Name:	
Call 1 Peer Phone:		Call 2 Peer Phone:	
Call 1 Duration:		Call 2 Duration:	
Call 1 Packets Sent:		Call 2 Packets Sent:	

Figure 5-2. The Info tab on the Sipura ATA's web interface

Table 5-2. Top 10 Sipura options

Tab title	Option name	Recommended value	Explanation
System	Primary NTP Server	*pool.ntp.org*	Sets the SPA's clock automatically.
System	Admin Password	Make it up!	Sets the admin password.
SIP	SIP T1	.5–2	Sets the SIP timeout value. Crank this up for high-latency network connections.

Table 5-2. Top 10 Sipura options (continued)

Tab title	Option name	Recommended value	Explanation
Regional	Time Zone	Your time zone	Sets the SPA's time zone.
System	Syslog Server	IP address of syslog server on your network	Very useful for debugging.
Provisioning	Upgrade Enable	Yes	Will use the URL from "Upgrade Rule" to upgrade the SPA's firmware automatically.
Provisioning	Provision Enable	Yes	This requires the Sipura profile compiler. If you have more than 10 SPAs, you should be able to obtain this tool to aid in configuration. Contact Sipura for more information.
X Line	RTP TOS/DiffServ Value	Varies	This controls the IP TOS value for Real-time Transport Protocol (RTP) or audio packets from the SPA. When used in conjunction with intelligent switches and routers, this can ensure excellent voice quality on your network.
X User	VMWI Ring Splash Len	0	This is a common request. It will disable the "splash ring" for voicemail notifications. Otherwise, your analog phone will chirp every so often if you have a voicemail. Very annoying.
X Line	Preferred Codec	Varies	Sets the preferred voice codec to use. Various codecs are available with quality/bandwidth trade-offs.

Dial-Plan Magic

Of all of the options on the Sipura, the *dial plan* lets you be the most creative. The dial plan is a string of characters that tell the Sipura how to treat calls—where to send them, any digits to add (or remove), etc. In its most basic use, the dial plan controls when to send calls.

VoIP devices are much like cell phones. You have to "send" the number as a whole to the remote server. But how does the ATA know when you are done entering digits? On the Sipura line of ATAs, this is controlled by two more parameters that you should be familiar with. They are called Interdigit Short Timer and Interdigit Long Timer, and you can find them on the Regional tab. Interdigit Short Time specifies the delay (in seconds) for sending numbers that match a string found in the dial plan. Interdigit Long Timer specifies the delay (also in seconds) for sending numbers that do not match the dial plan. Here is an example:

```
Line 1 Dial Plan: (7xxx)
Interdigit short time: 3 seconds
```

This means that when I dial 7104, the Sipura will send that number to the remote SIP server 3 seconds after I press 4. If I were to dial 2627638123, the Sipura would send that number to the remote SIP server 10 seconds after I entered 3 because there is no pattern matching that number. Let's take a look at a more complete example:

```
Line 1 Dial Plan:
([2-9]xx[2-9]xxx|[2-9]xx[2-9]xxxxxx|1[2-9]xx[2-9]xxxxxx|011[2-9].|7xx|7xxx)
Interdigit short time: 3 seconds
Interdigit long time: 10 seconds
```

This example matches NANPA 7-digit, 10-digit, and 11-digit dialing. It also includes NANPA international dialing, as well as matches for three- and four-digit extensions beginning with 7. This way, most standard dialing, as well as extension dialing, will be covered by this dial plan, thus matching the Interdigit Short Timer of 3. I should point out that if you want a number dialed immediately, regardless of whether it matches the dial plan, you can add # to the dial string. Thus, in the previous example, 12345678# will send 12345678 to the remote server immediately, even though it does not match the dial-plan string. It's probably worth pointing out that there is a limit to how long a dial-plan string can be. A dial-plan string has a maximum length of 2,047 characters. On the Sipura SPA-3000, you can have eight dial-plan strings for the Public Switched Telephone Network (PSTN) line. The limitation for those is 511 characters each.

Advanced Dial-Plan Examples

Here are some more-advanced dial-plan examples:

```
(<111:1002@192.168.0.22:5061)
```

This is a slight modification of the dial-plan string from "Build a Bat Phone" [Hack #63]. This string will call extension 1002 on the Sipura at 192.168.0.22 on port 5061. However, it will do this only if you dial 111. This is a very inexpensive way to set up a PBX with no SIP server at all. You could take several Sipuras with static IP addresses and assign them extensions. You

[HACK #63 Build a Bat Phone]

could even include an SPA-3000 for single-line POTS termination/origination. Here is a more complete version of the preceding code:

```
(<111:1002@192.168.0.11:5061>|<112:112@192.168.0.12:5061>|<113:113@192.168.
0.13:5061>)
```

If you had this same dial plan on every device, you would be able to call between them simply by dialing 111, 112, and 113.

This example is another slight modification. Essentially, here we are adding 1847 to any number that the user dials as seven digits:

```
(<:1847>[2-9]xx[2-9]xxx|1[2-9]xx[2-9]xxxxxx|011[2-9].|7xx|7xxx|xx.)
```

The following configuration will work on the SPA-3000 only:

```
([49]11<:@gw0>|*xx<:@gw0>|[2-9]xxxxxx<:@gw0>|[2-9]xx[2-9]xxxxxx<:@gw0>
|1800xxxxxxx<:@gw0>|18[6-8][6-8]xxxxxxx<:@gw0>|7xx|7xxx|1[2-9]xx[2-
9]xxxxxx|011[2-9].)
```

The following list explains what this dial plan does for you:

- Calls to 411 and 911 go to the PSTN via the POTS line.
- *xx (e.g., *69) goes out via POTS.
- Seven-digit and ten-digit calls go out via the POTS line.
- Toll-free calls go out via POTS.
- Three- and four-digit extensions are sent to the first SIP server defined.
- Eleven-digit long-distance numbers are sent to the first SIP server.
- International dialing is sent via SIP as well.

Here is yet one more advanced dial plan:

```
([49]11<:@gw0>|[2-9]xxxxxx|[2-9]xx[2-9]xxxxxx|1800xxxxxxx|18[6-8][6-
8]xxxxxxx|7xx|7xxx|1[2-9]xx[2-9]xxxxxx|<9:>xx.<:@gw0>)
```

This is very similar to the previous plan, however any calls prefixed with 9 that are longer than three digits will be sent via the POTS line.

These are limited examples of what you can do with the Sipura line of ATAs. After more experimentation, you will quickly realize how much fun you can have with a $70 ATA!

—*Kristian Kielhofner*

Build a Bat Phone
#63   Do you think Bruce Wayne uses VoIP to receive emergency calls from the Mayor of Gotham? Of course he does. He's *that* cool (his car is OK, too).

If you've worked your way through "Tweak Your Sipura ATA" [Hack #62], you know Sipura Technology makes some very powerful and flexible ATAs. So powerful, in fact, that you can use them to set up a point-to-point "hot line" with no SIP proxies or registrars.

A "bat phone" (or automatic ring-through in the telco world) is best known from the popular *Batman* television series. Batman would have such a burning desire to speak with the commissioner that he didn't even have time to dial. The simple act of picking up the phone automatically connected him to the designated remote station.

Here is what you will need to get this going with your two Sipuras:

Two Sipura ATAs

> As of this writing, the 841, 1000, 1001, 2100, 2000, and 3000 were widely available, but Sipura has just been acquired by Cisco, so these model numbers could change.

Static IP addresses or dynamic DNS

> Each Sipura will need to know where the other is. On a simple LAN, this is incredibly easy. Just assign static IP addresses, and move on. Over the Internet, behind NAT and/or firewalls, this task can get complicated. While it's too much to cover here, you will want to look into port forwarding and dynamic DNS.

First Things First

Take out one of your shiny new Sipuras. This will be called ATA1. Connect the phone (to line 1 if you have more than one line) and Ethernet cables. Then connect the power. If your LAN uses DHCP, the Sipura will acquire its IP address using DHCP. If you pick up your telephone, you should here a dial tone. Enter ********. You should hear a not-so-friendly voice say the words "Sipura configuration menu." At this point, you should enter **110#**. The same "friendly" voice should come back and read you your IP address. Make a note of it.

> While DHCP does make it easier to attach new devices, it makes it harder to keep track of them. Once you get into the web interface, you should assign a static address, or use the static mapping features of your DHCP server to assign the ATAs the same IP addresses at all times.

After you have made note of the IP address for ATA1, repeat the process for your other Sipura, ATA2. For the rest of this hack, we'll assume ATA1 and ATA2 have the respective IP addresses 192.168.1.101 and 192.168.1.102.

After you have the IP addresses of your Sipura devices, fire up a web browser on a machine connected to the same LAN. Using your web browser, enter the IP address of ATA1. You should see a gray screen filled with status information. Open another window (or tab) and enter the IP address of ATA2. You should see a similar (if not identical) screen, with the exception of the different IP addresses. Now we're ready to have some real fun!

Configure the Sipuras

The dial plan on the Sipura ATAs is one of the more attractive features of the SPA line of products [Hack #62]. It is the dial plan that is going to make this hack possible. In your web browser for ATA1, click on the Admin link in the top right-hand corner. You should see several more options become available. Then click Advanced. You should see even more options become available.

Next, click the Line 1 tab and scroll down to Username. Enter **ata1**. Do the same for Display Name. Scroll down to Dial Plan. In the Dial Plan edit box, erase what is currently there and replace it with the following:

```
(S0<:ata2@192.168.0.102:5060>)
```

Save your changes. Now, for ATA2. Switch over to the ATA2 browser window, and click Admin and Advanced again. Now, move over to the Line 1 tab, and down to Username and Display Name. Fill in **ata2** for both. Again, scroll down and fill the Dial Plan box, this time using the values for ATA1:

```
(S0<:ata1@192.168.0.101:5060>)
```

Again, save your changes. Now, any time you pick up either phone connected to line 1 on ATA1 or ATA2, it will automatically call the phone attached to line 1 on the other ATA.

Hacking the Hack

Nothing says "Holy phone mod, Batman!" like a bright-red rotary-dial phone with the mechanical dial wheel removed. Replicas of such phones are actually available on eBay, as are plans to build ones that have flashing lights, too. But for this hack, all you really need to do for an authentic red bat phone* is the following:

1. Find a cheap, old rotary-dial phone at a garage sale or in the attic.

2. Remove the dial wheel and discard it.

3. Carefully remove the electromechanical guts of the phone and set them aside.

---

* I've since learned that "true" bat phones not only are red, but must also reside in a glass cake cover when not in use.

4. Use some bright-red, plastic-friendly Krylon Fusion spray paint to turn that vintage monster red, like a tomato. Allow it to dry, of course.

5. Put the phone guts back into the newly blushing enclosure, reconnect to your Sipura bat phone ATAs, and use that hotline to your heart's content.

—Kristian Kielhofner

Brew Your Own Zaptel Interface Card
#64

With a little tweaking, a very common fax/modem card can become a clone of the single-line X100P interface card.

The Digium X100P foreign exchange office (FXO) card, used to connect a single phone company line to an Asterisk server, is actually an Intel V.92 Data/Fax/Voice modem card. One visual comparison between an Intel V.92 Winmodem PCI card and an official X100P, and it's obvious that the two cards are identical. So, using the less-expensive modem card in place of an X100P card is not only possible, it's downright easy.

The Intel 537EP chipset is a V.92 PCI modem chip family. Many modems are built on the Intel 537EP chipset, but this hack is known to work only with the Intel V.92 Winmodem card.

The critical thing about using an Intel V.92 modem card that has not been purchased from Digium as an X100P, but otherwise looks the same, is that the vendor ID encoded into the card will read differently, breaking the original Zaptel driver and rendering the card useless. Fortunately, there are two ways around this. The most obvious solution is to hack the code of the driver. Before you compile Asterisk and Zaptel from the Digium CVS archive [Hack #41], you'll need to edit the *zaptel/wcfxo.c* file.

Here's the existing code snippet you'll need to change:

```
static struct pci_device_id wcfxo_pci_tbl[] __devinitdata = {
{ 0xe159, 0x0001, 0x8085, PCI_ANY_ID, 0, 0, (unsigned long) &wcx101p },
{ 0x1057, 0x5608, PCI_ANY_ID, PCI_ANY_ID, 0, 0, (unsigned long) &wcx100$ };
```

Change this section in *zaptel/wsfxo.c* to this:

```
static struct pci_device_id wcfxo_pci_tbl[] __devinitdata = {
{ 0xe159, 0x0001, 0x8085, PCI_ANY_ID, 0, 0, (unsigned long) &wcx101p },
{ 0xe159, 0x0001, 0x8086, PCI_ANY_ID, 0, 0, (unsigned long) &wcx101p },
{ 0x1057, 0x5608, PCI_ANY_ID, PCI_ANY_ID, 0, 0, (unsigned long) &wcx100$ };
```

The line added in the middle will allow the Zaptel *wcfxo* driver to work with standard Intel V.92 Winmodem boards (while still keeping the driver compatible with official Digium X100P cards). Recompile the Zaptel drivers [Hack #41], and your Intel V.92 cards can be used as FXO interfaces. Pretty neat, eh?

The other, more difficult way to enable this feat is by modifying the boards themselves. This means re-creating the same modification that Digium does when it modifies Intel cards to create so-called genuine X100P cards. Remove the resistors marked R13 and R19 by unsoldering them. But be careful, and don't expect to return your Intel V.92 card, as its warranty will now be invalid.

Build a Speed-Dial Service on Cisco IP Phones

#65 HACK

Cisco's 7900 series IP phones have some powerful programmable firmware that you can harness for your own unique purpose, answering the age-old question, "Doesn't this LCD display seem a bit large for just caller ID?"

That expensive Cisco phone on your desk has some great hidden capabilities. Additional tools and toys that lurk beneath the 79xx's gray exterior could increase your productivity and foster some innocent fun. I'm talking about things like automated weather reports on the phone's display, simple menu-driven applications (like a time card, say), and just about anything else you can program using an XML web site.

In fact, you'll probably build so many cool tools and toys that you'll need a way to sort through them, like a directory or a menu. Using the Cisco VoIP phone's XML application capability, you can set up such a directory.

For this hack, you're going to use XML to create your own custom menus that access hidden features of the Cisco phone. To make your menu appear on the phone, you'll need to configure the phone to look for your custom menu file.

Cisco phones, like most IP phones, have a Flash storage device onboard that is checked and optionally updated at every startup. During startup, the phones contact a TFTP server and attempt to download settings stored in files on the server [Hack #80]. By setting the services_url property on the phone's console to the URL of your menu, you can configure the phone to load your custom menu upon its next reboot. To set this property, press the Settings button on the Cisco phone. Then, select Unlock Config and use the keypad to enter the password (the default is cisco). Next, select the XML URL option, followed by the Services URL option. Enter the URL, like the following example, using the phone's keypad (you enter slashes and colons by cycling through the # and * keys):

```
http://lathama.com/services/cisco/
```

Now, when the phone boots, it will attempt to load a menu for the Services button by accessing an XML file at the URL. You can also set this URL in a configuration file on the TFTP server. (If you do neither, the Services button will be rather useless, as it doesn't default to any built-in settings.) Edit the *SIPDefaults.cnf* file on the TFTP server to set the Services URL. Use an entry like this:

```
services_url: "http://lathama.com/services/cisco/"
```

Leave the ending slash in there. It will tell the web server to grab the default or index file. Testing this with a web browser will be difficult, as the XML has no headers to tell the browser what is going on. Create a file called *index.html* (*default.htm* for Windows web servers) on a nearby web server that's accessible at the URL you supplied to the phone, and fill it with XML content like this:

```
<CiscoIPPhoneMenu>
 <Title>My New Menu</Title>
 <Prompt>Prompt This</Prompt>

<MenuItem>
 <Name>Menu Item</Name>
 <URL>http://domain/cisco/services/menuitem.xml</URL>
</MenuItem>

<SoftKeyItem>
 <Name>Soft Key</Name>
 <URL>http://domain/cisco/services/softkey.xml</URL>
 <Postition>1</Posititon>
</SoftKeyItem>

</CiscoIPPhoneMenu>
```

This sample shows you a simple yet slick method of loading a menu. When you press the Settings button on the phone, the menu shows up as simple text bars and allows you to arrow-down or up to the option you choose, using the phone's navigation controls and softkeys. After you make a selection, the listed URL is queried. This convention is the formula behind what Cisco calls *services*. You can think of services as the little applications (stock tickers, games, etc.) on modern cellular phones. The Cisco phones split the phone directory off as its own service (available through the Directory button rather than the Services button), but that shouldn't stop you from making your own directory with this hack.

The menus are simple to create, but you'll need to become familiar with the strange tags. If you're not familiar with XML, the `<CiscoIPPhoneMenu>` tag, which is used to enclose Cisco phone menu structures, might seem a little strange to work with. I would have just used `<Menu>`, but that might not work in newer firmware. Menus contain one or more menu items denoted with the tag `<MenuItem>`. A softkey type of menu that uses the `<SoftKeyItem>` tag also is available. Softkey menus use the buttons on the side of some of the phones (the 7960, for example, has six on the right and four at the bottom of the display). Some of the older Cisco phones do not have the extra buttons, so for compatibility with all of Cisco's IP phones, stick to simple menus.

Inside <MenuItem> are easy tags, such as <Name> and <URL>. These are case-sensitive on some current firmware versions. <Name> is what shows up as the item's name, and <URL> is what it will request if the user selects the menu option. Notice that the http:// part is in the URL, but FTP URLs will work, too.

Now, on to your first menu. Start the following example and adjust it to suit. Let's say you use the extension of the phone for the title, to allow you and your friends or users to know the extension at any time. Break up the menu into chunks that you can handle:

```
<CiscoIPPhoneMenu>
 <Title>EXT 1234</Title>

<MenuItem>
 <Name>PIM</Name>
 <URL>http://domain/cisco/services/pim.xml</URL>
</MenuItem>

<MenuItem>
 <Name>Work</Name>
 <URL>http://domain/cisco/services/work.xml</URL>
</MenuItem>

<MenuItem>
 <Name>Play</Name>
 <URL>http://domain/cisco/services/play.xml</URL>
</MenuItem>

</CiscoIPPhoneMenu>
```

So you now have three sections to add things into. Work on PIM first, and add a list of favorite restaurants (hey, you'll want to reward yourself with a pizza when this is through!). Go to the *pim.xml* file on the web server. Add some local and not-so-local places, and then allow the menu to dial the number for you. To keep my favorite places safe, I'll suggest fictional restaurants and phone numbers. (Actually, if you're in the Detroit area, nothing beats National Coney Island!)

```
<CiscoIPPhoneMenu>
 <Title>EXT 1234 - PIM</Title>

<MenuItem>
 <Name>Restaurants</Name>
 <URL>http://domain/cisco/services/pim.xml</URL>
```

```
  </MenuItem>

  </CiscoIPPhoneMenu>

  <CiscoIPPhoneDirectory>
   <Title>EXT 1234 - PIM - Restaurants</Title>

  <DirectoryEntry>
   <Name>Larrys Latkes</Name>
   <Telephone>18005551212</Telephone>
  </DirectoryEntry >

  <DirectoryEntry >
   <Name>Timbos Nacho World</Name>
   <Telephone>18665551212</Telephone>
  </DirectoryEntry >

  <DirectoryEntry >
   <Name>Drunken Cow Steak House</Name>
   <Telephone>18775551212</Telephone>
  </DirectoryEntry >

  </CiscoIPPhoneDirectory>
```

Look at all that's changed. You are now using the `<CiscoIPPhoneDirectory>`
tag to show the dial and other buttons at the bottom of the screen. You also
are now using `<DirectoryEntry>` rather than `<MenuItem>`, and `<Telephone>`
rather than `<URL>`. To complete the hack, edit the *work.xml* and *play.xml*
files.

—Andrew Latham

HACK #66 Power Cisco Phones with Standard Inline Power

To avoid lock-in with Cisco-only phones and switches, learn how to power
Cisco phones from non-Cisco switches.

IP phones can be powered through their Ethernet connections. The standard for this inline power is called *802.3af*, and many equipment manufacturers support it—except for Cisco, which uses its own proprietary inline
power method. Because of this, you can match Cisco IP phones only with
Cisco-powered switches (unless you use Cisco's only phone model to support 802.3af—the 7970). This is an unfortunate form of vendor lock-in, but
all is not lost. You can do a couple of things to get Cisco IP phones to draw
power from non-Cisco switches.

If your budget permits, the obvious (though proprietary) solution to this
problem is to use Cisco PoE switches to power the phones. Some other
switch makers, like Foundry Networks, also support Cisco's proprietary PoE

standard. If you can't afford to forklift your switches, you might instead want to power your Cisco phones by way of a *power injector*, which is a patch panel that adds inline power to a CAT5/CAT6 cable connection. Consider Cisco PoE-compatible injectors like those made by PowerDsine (*http://www.powerdsine.com/*).

But, if you can't do that either, do the next best thing: *hack*.

> Hacking inline power will almost certainly void your IP phone's warranty, and probably your switch's or power injector's, too. A short circuit could fry your switch *and* phone if you're not careful. Proceed with caution!

By changing some wires on a standard UTP Ethernet patch cable, you can make a compatibility cable that lets you plug Cisco IP phones into any 802.3af source, as shown in Figure 5-3. Essentially, you are flipping wires 4 and 7, and 5 and 8. Be advised, this technique could void the warranty of your phone and your switch.

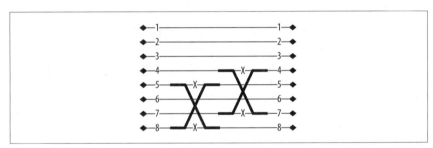

Figure 5-3. The wiring diagram for a hacked PoE cable

Make sure your switch lets you program, port by port, which ports get power and which ones don't, because in a native Cisco PoE solution, Cisco IP phone power requirements are "auto-detected," so power can turn itself on and off as necessary on each port. There's no such provision when using a hacked cable to supply 802.3af power to a Cisco PoE-using phone. If this is a problem, and 802.3af won't work with the hacked cable, try using a device that does the two-pair flip but also works with auto-detection, such as 3Com's 48-volt IntelliJack switch converter, part number 3CNJVOIP-CPOD.

> The Cisco 7970 IP phone *does* support 802.3af power sources, unlike the more popular (and less expensive) 7960 and 7940 phones.

Customize Your Cisco IP Phone's Boot Logo

Change the logo on your Cisco IP phone, and reflect your inner geek's refined sense of monochrome style.

Have you ever wished you could change the boot-up logo on your cell phone? Have you ever wanted to use custom graphics on your appliances' LCD screens? Most Linux geeks love to plaster Tux the Penguin, the official mascot of Linux, all over the place—and what better place than a hackable display? If you're like me and you have a thing for the penguin, allow our underdressed friend to show himself on your Cisco VoIP phones.

First, the facts: when most IP phones boot, they look for configuration files on a nearby TFTP server and download them to configure the phone further [Hack #80]. The configuration files allow the specification of a logo, along with other tweakable goodies. By editing or adding the logo_url setting in a phone's configuration file, you can dictate which logo the phone should use. The storage or location for this logo varies depending on the version of the firmware that's loaded on your phone, but you should be able to specify a standard HTTP URL to point the phone toward its logo. (This kind of hack is also possible on other phones, so look it up.) Here's the specific setting for a Cisco configuration file:

```
logo_url: "http://domain/cisco/logo.bmp"
```

As you can see, the URL points to a bitmap picture (logo.bmp). So far, this looks to be a simple hack, and with a few notes, it will stay that way. The image that the phone downloads is on a server, so it needs to find the server somehow. In other words, make sure the phone's DNS server setting is right so that it can resolve the hostname you provide in place of the *domain* placeholder in the URL.

The image size and color are also important. If you use an image with the wrong size or aspect, it will come out looking a bit funky on your Cisco's LCD. The default size for the display on the Cisco 7960 is 133×65 pixels, so that would be a good place to start. The image should be monochrome, at least for the 7960. Color is supported on newer models, like the 7970G. (You can always specify a different URL in your color phone's config files that points to a color version of the same image if you need to support color and monochrome displays.) The older and current phones will alter an image to fit the screen and color if it does not match. This auto-correction might not be perfect, so you might want to run your image through Photoshop or the GNU Image Manipulation Program (GIMP) to meet the size and color requirements.

When I did this hack, I used a PNG-format picture of Tux the Penguin. Tons of great images like this are available at *http://images.google.com/*. I opened my image in the GIMP to have a look. Then I resized the happy fellow so that his height matched the height of the LCD, 65 pixels. Finally, I converted to grayscale and saved. Figure 5-4 shows the finished product. Cute, isn't he?

Figure 5-4. The Tux logo, as he appears on a Cisco 7960's display

I then simply uploaded the file to my web server, at the URL specified in my 7960's TFTP configuration file. The next time I booted up my 7960, there was Tux, happy as usual.

—Andrew Latham

Configure Multiple IP Phones at One Time

Uniden's IP phones, like the UIP200, are excellent business-grade telephones that can be mass configured by TFTP—that is, if you know how.

As you might have already gathered, you can configure IP phones in three ways: directly, using the phone's LCD and buttons; through a web or Telnet interface; or via a configuration file the phone downloads from a TFTP server during boot-up. Those three methods are presented in order of the level of detail to which you can administer the phone, with TFTP configuration allowing the most precise control. Once you've got a TFTP server running [Hack #80], you need only drop the right text files onto your TFTP server to mass-configure your IP phones.

The Uniden IP phones don't offer the trademark high-end look and feel of the Cisco 7900 series of phones, but they do provide a good value nonetheless. At half the price of a typical Cisco SIP phone, the Uniden UIP200 SIP phone supports all of the standards fundamental to a VoIP LAN: inline power, SIP, and several of the key audio codecs. Getting a UIP200 onto the network and doing useful things with your softPBX is a snap, with the help of mass configuration via TFTP.

Get the Uniden on the Network

To get started, I'll assume your UIP200 is connected to your Ethernet LAN and is powered up. You might want to use a static IP configuration on the phone (as opposed to DHCP), so pop into the Uniden's Quick Setup utility by pressing the phone's Menu key. Use the directional arrow keys to access the Network Settings menu. Then, press the Menu key to select it. Now, press the down arrow until you reach IP Address; then press Menu again. Now, you can key in the IP address, substituting stars (*) for periods. Press Menu to accept the address, and then arrow down to the Subnet Mask and enter the appropriate value for your LAN. Repeat this process for the Default GW option, entering the right value for your LAN's default gateway (probably the address of the nearest router).

Next, use the arrow keys to select DNS Server 1, press Menu, and enter the IP address of the nearest DNS server. Then, reboot the phone by powering it off and on. Try pinging the phone's IP address from a nearby PC to see if it's communicating with the network.

Connect the Uniden to TFTP

To alter the phone's automatic configuration mode (i.e., TFTP-based configuration), you need to unlock the configuration menu. To do so, press the phone's Menu key and press the down arrow until you reach Unlock Config. Press Menu again. You'll be prompted for a password, which you can enter using the number pad (on most UIP200 firmware versions, the password is 2002). Press the Menu key to confirm the entered password.

Next, press the up arrow until you reach Network Settings. Press Menu; then press the up arrow until you reach TFTP IP Address. Press Menu, enter the address of your TFTP server, and then press Menu again to confirm it. Now, press the Cancel key to go back to the main menu. Use the arrows to find Phone Settings, and press Menu. Then, it's the arrows again until you find Auto Config. Press Menu, and then press the up arrow to set Auto Config to Enabled. Press Menu, press Cancel, and then press Menu twice to reboot the phone. On the next boot, the phone will look to your IP address to find its SIP and telephony configuration.

Build a Uniden Configuration File

The best way to learn Uniden SIP configuration is to step through the Uniden configuration files. But before we do so, let me go over the basic structure a Uniden UIP200 phone looks for when booting up and searching for its config on a TFTP server. First, the phone expects to find a file called *unidencom.txt*, which describes the operational characteristics that apply to all phones—things like site-wide audio settings, addresses of SIP proxies (softPBX servers, that is), DNS servers, and the like. The phone will also look for a file called *uniden<MAC>.txt*, where *<MAC>* is the MAC hardware address of the phone. So there is a single *unidencom.txt* file, and there are many *uniden<MAC>.txt* files.

I'll step you through the key settings found in *unidencom.txt*:

```
ProxyServer                10.1.1.10
ProxyServerPort            0
OutboundProxy1             10.1.1.10
OutboundProxy1Port         0
```

The proxy settings tell the phone which SIP server to deal with when resolving dialed numbers and attempting to connect calls. `ProxyServerPort` allows you to override the default SIP UDP port of 5060, if your softPBX is configured this way (not likely). `OutboundProxy1` allows you to specify that you want to use a different SIP server for nonlocal calls. In most cases, `OutboundProxy1` will be the same as `ProxyServer`.

```
Registrar1                 10.1.1.10
Registrar1Port             10.1.1.10
```

Like a SIP Proxy, which routes calls, a SIP Registrar also handles connections from SIP phones, but for a different reason. A SIP Registrar keeps tabs on SIP users and informs requesting callers as to where a particular SIP user can be found (i.e., what IP address that user is registered from). You can specify a different registrar. But in most cases, this will also be the same address as your SIP Proxy, as proxies and registrars tend to be on the same server more often than not.

```
DnsServer_1                10.1.1.10
```

This setting lets you override whatever DNS server address you provided in the Quick Setup menu—unless the phone is using DHCP, in which case it will use whatever DNS server is recommended by the DHCP server when it acquires its address.

```
RegisterExpireSec          3600
RegisterRetrySec           90
```

These settings tell the phone how often to register with the SIP registrar, and how often to retry failed registrations.

```
SipPort                    5060
```

`SipPort` tells the phone which UDP to use when listening for SIP messages, such as incoming calls. 5060 is the default, and in most scenarios, you want to leave it at 5060.

```
G711MuTxPacketLength       20
G711MuJitterBufferLength   10
G711ATxPacketLength        20
G711AJitterBufferLength    10
G729TxPacketLength         20
G729JitterBufferLength     10
```

The Uniden's three supported codecs are G.711 muLaw, G.711 aLaw, and G.729. The G.711 codecs are standard 64 Kbps PCM bitstream codecs that mimic the sound framing technology used by the legacy time division multiplexing (TDM) equipment on the public telephone network. G.729 is a high-compression codec that requires about half the bandwidth of G.711. These settings allow you to tweak the packet length (in milliseconds) of each codec. Adjusting the packet length (also called the *packet interval*) changes how large each sound packet will be. Shorter lengths will yield smaller packets, but will require greater bandwidth because they incur more Ethernet and IP overhead. (For a great description of how packet intervals and overhead interact, if I do say so myself, pick up O'Reilly's *Switching to VoIP*.)

The jitter buffer settings tell the UIP200 how many milliseconds of sound data to record before transmitting to overcome the commonplace network instability known as jitter. In all reality, you might not need to touch any of these settings, though it is certainly fun to toy around with the jitter buffer length if your wide area network (WAN) link is particularly jittery.

```
DiffServMode                OFF
DefaultDiffServParam        192
RTPDiffServParam            160
```

DiffServ is a Quality of Service (QoS) mechanism that uses a policy-based approach to enforcing different classes of service on the same WAN. It's not a bad idea to switch DiffServMode to ON, but don't expect this to increase the quality of your phone calls over the Internet, as most Internet routers don't support the DiffServ standard. Class of service is useful only in a controlled enterprise environment.

```
VlanMode                    Disable
VlanID                      1
PcVlanID                    2
```

Virtual LANs, or VLANs, are a way Ethernet switches can segment traffic to create logically independent networks on the same equipment. In most enterprise VoIP scenarios, there are separate VLANs for voice and data traffic. These settings allow you to specify which VLAN ID to join the phone with. And since the phone has a built-in Ethernet switch for piggyback connection of a PC, you can specify the PC's VLAN, too. That way, even though phone and PC are connected on a single uplink cable, they can still be on separate VLANs. This functionality is common on most IP phones that double as Ethernet switches. Remember that since we're talking about *unidencom.txt*, the VLANs specified here will apply to all phones that use this TFTP server.

```
TftpAddress                 10.1.1.10
```

Just in case the DHCP server ever crashes and the phone can't acquire a TFTP server address from it, you can set the address of the TFTP server here.

```
TimeZone              -6
EnableDST             YES
EnableSNTP            YES
SntpServerIP          10.1.1.10
SntpRetrySec          1800
```

These settings control the time and date configuration on the phone. `TimeZone` sets the local time zone of the phone, expressed as an offset of Greenwich Mean Time. In this case, the phone is 6 hours behind because it's in the Eastern time zone. `EnableDST` allows the phone to switch from Eastern Daylight Time to Eastern Standard Time automatically, and vice versa, and `SntpServerIP` and `SntpRetrySec` enable the phone to use Simple Network Time Protocol to synchronize its clock with the other devices on the network.

```
PreferredCodec        g711u,g711a,g729
Language              English
```

The `PreferredCodec` setting tells the phone which codecs you prefer to use when connecting calls. If the device on the other end of the call is deemed not to support your preferred codec, the phone goes to the next one in the list. G.711 uLaw is most common in the United States, and G.711 aLaw is common elsewhere in the world. G.729 is a bandwidth-conserving codec. If this phone is going to be used by a road warrior or telecommuter with unpredictable bandwidth capacity, or used over a small (128 kbps at the least) wide area link, you might consider putting G.729 at the front of this list. Uniden's firmware doesn't like spaces in this list, by the way.

`Language` tells the phone which language to use for the menu prompts. Your choices are English, Spanish, and French (sorry, *übergeeks*: no Klingon).

```
StunServerAddr           0.0.0.0
StunServerPort           3478
```

Simple Traversal of UDP NATs (STUN) is a protocol that helps IP phones deal with the problem of Network Address Translation (NAT), a common technique employed by many firewalls to mask a group of privately addressed devices (like IP phones) behind one or more public IP addresses. The protocol is dealt with in more detail in Chapter 6. Leaving `StunServerAddr` at `0.0.0.0` disables the UIP200's STUN client, and `StunServerPort` allows you to override the default port number.

```
DirectIpDialing          No
```

To enable direct dialing by IP address (so that you can call another IP phone by its address rather than its phone number), as shown in the example given in "Make IP-to-IP Phone Calls with a Grandstream BudgeTone" [Hack #60], change this setting to Yes.

```
AdminPassword                    1234/5678
```

The `AdminPassword` setting allows you to change the menu password rapidly on all of the phones that get their configs from this TFTP server. The format is *oldpassword/newpassword*.

HACK #69 Customize Uniden IP Phones from TFTP

Use unique configurations on each IP phone, and while you're at it, do some firmware revision control, too.

There are two files for each Uniden IP phone on the TFTP server: one that's shared by all of the phones on the network (*unidencom.txt*, described in "Configure Multiple IP Phones at One Time" [Hack #68]) and one that's exclusive to each phone on the network. These exclusive, phone-specific config files, whose filenames contain the names of their corresponding phone's MAC hardware address, control the firmware and hotkey setup of that particular IP phone. I'll step you through a sample Uniden phone-specific config file, as it might appear on your TFTP server:

```
AutoFirmwareUpdate               YES
FirmwareFileName                 uip200_455enc.pac
FirmwareVersion                  BS4.55
```

Enabling `AutoFirmwareUpdate` with a `YES` will cause the phone to attempt a firmware patch automatically when it boots. It will try to grab (and install) the firmware package specified by `FirmwareFileName` from the TFTP server. The desired firmware version is specified by `FirmwareVersion`, and the phone will grab the firmware file you specify only if the version is different from the version currently running on the phone.

```
MyLcdDisplay                     Maddie's Phone
MyDialNumber                     1138
DisplayName                      Madelyn
UserNameForProxy                 1138
PasswordForProxy                 uniden
UserNameForRegistrar             1138
PasswordForRegistrar             uniden
```

`MyLcdDisplay` determines what greeting to display on the phone when it is waiting to call or be called, and `MyDialNumber` determines what number to display. `DisplayName` attempts to set the caller ID name to be used on outgoing calls, if the softPBX supports this. `UserNameForProxy`, `PasswordForProxy`, `UserNameForRegistrar`, and `PasswordForRegistrar` establish the login credentials to be used when the phone logs into the SIP servers that handle its calls (proxies and registrars are often hosted on the same server, so the credentials are often identical).

```
ProgrammableKey1        OneTouchDial
ProgrammableKey2        TwoTouchDial
ProgrammableKey3        CallForward
ProgrammableKey4        DoNotDisturb
ProgrammableKey5        VMA
ProgrammableKey6        Mute
```

The `ProgrammableKey1` through `ProgrammableKey8` settings allow you to assign functions to the UIP200's hotkey. Here's what the possible values do:

OneTouchDial

Causes the phone to dial a phone number (supplied in the `OneTouchKey` settings later in the file).

TwoTouchDial

Causes the phone to dial a phone number that's associated with one of the 10 digit keys on the dial pad (these 10 numbers are supplied later in the file).

CallForward

Enables call forwarding, if supported by the softPBX to which the phone is connected.

DoNotDisturb

Causes the phone not to ring, even when calls are received (if voicemail is available courtesy of your SIP proxy, it will answer calls instead; otherwise, the calling party gets a busy signal).

VMA

Voice Mail Access. Causes the phone to dial a number associated with retrieving voicemail messages. The exact number is specified later in the file.

Mute

This is a standard telephone mute setting that disables the microphone in the phone so that it won't pick up input on your end.

```
OneTouchKey1        18005551212
OneTouchKey2        411
TwoTouchDigit0      3000
TwoTouchDigit1      3001
TwoTouchDigit2      3002
```

You use the `OneTouchKey1` through `OneTouchKey4` settings to supply the phone numbers that are used with up to four `ProgrammableKey` settings, so you can set up to four of the UIP200's eight hotkeys to be one-touch dialing keys. The `TwoTouchDigit0` through `TwoTouchDigit9` keys, on the other hand, are used to set up two-touch dialing (first the hotkey, and then a number key on the dial pad). Values supplied here become the phone numbers that are called whenever a two-touch dial occurs.

```
VmaDirectCallNo    8080
VmwiLampIndicator  Enable
```

VmaDirectCallNo tells the phone what number to call when the VMA hotkey is pressed. VmwiLampIndicator, when Enabled, permits the phone to light its message-waiting indicator light. It's probably not a good idea to disable this one.

Once you've got this file set up the way you like, save it in the format *uniden<MAC>.txt*, where *<MAC>* is the Ethernet hardware address of the phone it applies to. Then, reboot the phone!

HACK #70 Control the Lights Using Your IP Phone

Using an X10 phone controller, you can turn your lights on and off from the comfort of your IP phone.

X10 home-control interface equipment has been a favorite pastime of geeks for decades. Since the early days of 8-bit hobby computers, you've been able to automate your home using your keyboard (and later, your mouse). X10 interface controllers connect to lights and other appliances in your house, and your computer can send serial commands to the controllers to turn them on and off and adjust voltage like a dimmer. Some X10 interfaces even offer telephone-based user interfaces, letting you control them by calling them with your phone.

There are a few ways to integrate X10 controls with an Asterisk phone system. The integration method depends on the type of X10 controller purchased. For this hack, I chose the X10 TR16A phone controller that operates by DTMF digits. Ordinarily, you would hook a phone line to it and then call that phone line with a standard phone to operate the TR16A. But, with Asterisk, you can connect directly to the TR16A as if you yourself are the phone company. Then, controlling the TR16A is as simple as a dial-plan hack in Asterisk.

The Asterisk system I used to connect to the TR16A contains a single Wildcard TDM400P using two foreign exchange station (FXS) modules and one FXO module. The system has one analog phone and one SIP-enabled Polycom IP500 phone. The SIP phone is the one I used as my "remote control"—the phone from which I sent my commands to the X10 controller.

I performed the following steps to integrate the TR16A with the Asterisk system.

For the TR16A controller:

1. I connected it to a suitable 120VAC outlet.

2. I set the Answer delay switch to Minimum.

3. I set the controller to the appropriate house code.

4. I set the four-digit PIN to verify access when accessing the unit. (The device includes instructions on how to do this.)

5. I used a modular cord to connect from the RJ11 port to the Asterisk FXS port.

For the Asterisk system:

1. I assigned the FXS port an extension number using an Asterisk `Dial` command.

2. I set the SIP peer for the IP phone to pass digits in-band (`dtmfmode=inband`).

3. I set the SIP peer for the IP phone to use the G.711 uLaw codec.

Here's the bit from */etc/asterisk/extensions.conf* that you would use, assuming the FXS port is `Zap/1`:

```
exten => 100,1,Dial(Zap/1)
```

Here's the bit from */etc/asterisk/sip.conf*:

```
[200]
username=200
secret=200
type=friend
dtmfmode=inband
disallow=all
allow=ulaw
```

Verifying if the setup is correct requires a few test calls. Use the Polycom IP phone to call the controller. Go off-hook on the IP phone and dial the extension number assigned to the controller (100 in this example). The line will ring for about 15 seconds before the controller will answer. When the controller answers, you will hear three beeps. Enter your PIN code, and then a second set of three beeps will confirm that you entered the proper PIN code. After the second set of three beeps, enter the module number you want to control, followed by * or # to turn the module on or off, respectively.

I mentioned that there are a few ways to integrate X10 controls with Asterisk. Using the TR16A, I demonstrated how you could connect to an Asterisk phone system using an FXS port and an IP phone to operate the X10 controller. X10 has another controller known as the CM15A. The CM15A connects to your local PC using a USB cable, and it boasts a free SDK that you can use to write custom scripts. Using the Asterisk system's IVR function, you can program Asterisk to run a script. In turn, that script can control the X10 modules, allowing for a pure software-based solution—that is, no FXS interfacing necessary.

Hacking the Hack

To have Asterisk dial the PIN code for you automatically, add it to the extension definition that dials the controller (assuming your PIN is 1212), so that it dials your pin automatically after connecting:

```
exten => 100,1,Dial(Zap/1,30,D(1212))
```

—Joel Sisko

Use a Rotary-Dial Phone with VoIP

HACK #71

Or "How Grandma Mabel Learned to Love Voice over IP."

What do a 19th-century Missouri undertaker and SIP have in common? The common thread is their ability to connect two parties without a third party intervening. That is, they both serve an intermediary role between caller and callee, signaling the call and providing a pathway for its sound signals.

In 1891, a Missouri undertaker, Almon B. Strowger, was granted a patent for an electromechanical device called a *stepper switch*. The stepper switch allowed the calling party to control whom he would connect to without the need for an operator. Today, that control mechanism might be analogized to a phone number. But the important point is that Strowger's invention made it possible to connect phone calls without a telephone operator's intervention.

SIP promotes the ability to call a party without the need for a SIP gateway or gatekeeper, as long as you know the recipient's *sip@* address, called a SIP Uniform Resource Indicator (URI). To pay homage to the pioneering Mr. Strowger, we have included this hack to connect a rotary phone, such as a 1920 Western Electric candlestick phone (Figure 5-5), to a VoIP network.

First, let me give you a quick overview of how a rotary phone works. The rotary phone provides signaling to the central office by establishing a flow of current and then interrupting the flow momentarily to signal a pulse. The number of pulses produced in a given time period represents the number being dialed. Five pulses represent the number 5.

Adding a rotary phone to a VoIP network is a relatively painless process, as long as you can build an intermediary gateway that understands pulse signaling and SIP, and can signal both legs of the call. What *can* be painful is finding a VoIP device that supports rotary or "pulse" dialing. That's why you might be better off building this gateway yourself.

Do Pulse with an IAXy

The Digium IAXy FXS gateway supports pulse dialing, providing a simple and complete solution. In fact, the IAXy is a complete gateway in and of

Figure 5-5. A 1920s vintage Western Electric candlestick phone

itself: it has an analog phone port on one side and an Ethernet RJ45 port on the other, complete with a little Inter-Asterisk Exchange (IAX) user agent built in.

The Digium IAXy supports pulse dialing out of the box. Configure */etc/ asterisk/iax.conf* with some entries that will allow the IAXy to be used as a peer:

```
[rotary]
type=peer
username=asterisk
secret=supersecret
```

In this case, I used rotary as the IAX peer name. All you need to do, once this config is done, is register the IAXy with your Asterisk server as the IAX user *asterisk*. The rotary phone connects to the IAXy. Now you can use the rotary phone as you would any valid endpoint in your dial plan by routing calls to IAX/rotary using a Dial command.

Do Pulse with a Wildcard

If you opt to use Digium's Wildcard TDM400 with an FXO module installed, you'll need to add a line of code to the */etc/asterisk/zapata.conf* file on the Asterisk phone system. In the Zaptel channel section of *zapata.conf* for this particular FXO channel, the following bit of config will enable the FXO port to support pulse dialing:

```
pulsedial=yes
```

Once you have modified the *zapata.conf* file, connect your home line to the FXO port. To test, go off-hook and try dialing an extension off the Asterisk system from the rotary phone. Remember that using a rotary phone takes a little longer to dial, so be patient. Once you have confirmed that you can dial internally, access the outside line configured for pulse dialing and call someone.

"Pass Through" Pulse Dialing Signals

With many of the other VoIP product manufacturers, the rotary end-to-end solution is not as easy to configure or operate. Most VoIP gateways support pulse dialing on the FXO connections to the phone company. But they typically do not support pulse dialing with the FXS connections (i.e., the phones). A simple hack to overcome this limitation, at least in theory, is available on some of these gateways. One could assign a dedicated FXO port that is enabled for pulse dialing directly to an FXS port. The FXS port in turn would automatically seize the outside line when the rotary phone goes off-hook. In this way, the pulse signals sent by the phone are "passed through" to the phone line.

The drawback of this configuration is that you will not be able to dial internal extensions directly. Oh, and you'll be scratching your head if your local central office doesn't support pulse dialing! Make sure your local central office supports pulse dialing, or this hack definitely won't work.

Do Pulse Without Any Special Hardware

If none of the previous solutions turns your crank, there's yet another way you can use your classic rotary-dial phone with modern telephony services like VoIP and your tone-dialing-only phone company. If you have Yahoo! Widgets (formerly Konfabulator), you can actually do tone dialing via your rotary phone with the help of your Windows PC or Mac. Download and install Harry Whitfield's totally 'leet DTMF Dial widget (*http://www.widgetgallery.com/view.php?widget=35922*).

By taking your old-school phone off the hook and holding its mouthpiece up to your computer's speakers while dialing on the widget, the DTMF tones this widget generates will be sent through the phone line to the phone company (or to your VoIP ATA if you use a VoIP service). Just make sure the volume on your computer is up high enough for the mouthpiece to pick up the sound of the tones.

—Joel Sisko

CHAPTER SIX

Navigate the VoIP Network
Hacks 72–87

Switching to Voice over IP—especially in an enterprise environment—is wrought with perils that you won't experience on a non-VoIP network. Real-time applications like voice require a high-quality, real-time network. And, at least by itself, traditional Internet Protocol (IP) networking gear doesn't fully deliver on that promise. Fortunately, you can apply some old tools—such as Perl and Ethereal—to VoIP networking to troubleshoot and improve your IP network.

When problems occur, your trusty old network troubleshooting apps will come to the rescue. In this chapter, you'll use Ethereal to sniff Session Initiation Protocol (SIP) signaling messages as they traverse the network inside of User Datagram Protocol (UDP) packets. If you're a seasoned hacker or a timid script kiddie, you can start with this chapter's Perl scripts, which graph and monitor VoIP activity on the network. You'll also be able to monitor latency and jitter—the two things VoIP admins want to avoid like the plague—using standard IP networking commands.

If you play your cards right, you might even learn how to beat a SIP-mangling firewall. (OK, you don't have to play cards at all. Just read "Explore NAT Traversal" [Hack #76].) I'll also show you how to clandestinely monitor and record actual VoIP phone calls—the IP equivalent of a phone tap.

 Monitor VoIP Devices

#72 The only thing worse than having a VoIP service outage is being the last to know about it.

It's the phone company's job to monitor traditional telephony links. Some legacy phone vendors, such as Avaya, even monitor their PBXs in the field via phone links back to their support headquarters. But no such convenience exists for downtime-wary VoIP administrators. Thanks to Perl, though, good system monitoring for VoIP is within your grasp. In this hack, you'll develop a Perl script that monitors SIP hosts over the network and reports back on their availability.

This script determines that the SIP host is alive by sending a SIP OPTIONS packet to the remote host and receiving a response. This determines not only whether the host is reachable via the network, but also whether the SIP application on the other end is listening and responding to requests.

The Code

Your Perl development environment will need the Time::HiRes module for this hack. Grab it from *http://search.cpan.org/dist/Time-HiRes/*:

```perl
#!/usr/bin/perl
use IO::Socket;
use POSIX 'strftime';
use Time::HiRes qw(gettimeofday tv_interval);
use Getopt::Long;
use strict;

my $USAGE = "Usage: sip_ping.pl [-v] [-t] [-s <src_host>] [-p <src_port]
<hostname>";

my $RECV_TIMEOUT = 5; # how long in seconds to wait for a response

my $sock = IO::Socket::INET->new(Proto => 'udp',
                                 LocalPort=>'6655',
                                 ReuseAddr=>1)
    or die "Could not make socket: $@";

# options
my ($verbose, $host, $my_ip, $my_port, $time);
GetOptions("verbose|v" => \$verbose,
           "source-ip|s=s" => \$my_ip,
           "source-port|p=n"=> \$my_port,
           "time|t" => \$time) or die "Invalid options:\n\n$USAGE\n";

# figure out who to ping
my $host = shift(@ARGV) or die $USAGE;
my $dst_addr = inet_aton($host) or die "Could not find host: $host";
my $dst_ip = inet_ntoa($dst_addr);
my $portaddr = sockaddr_in(5060, $dst_addr);

# figure out who we are
$my_ip = "127.0.0.1" unless defined($my_ip);
$my_port = "6655" unless defined($my_port);

# callid is just 32 random hex chars
my $callid = ""; $callid .= ('0'..'9', "a".."f")[int(rand(16))] for 1 .. 32;
# today's date
```

```
my $date = strftime('%a, %e %B %Y %I:%M:%S %Z',localtime());
# branch id - see rfc3261 for more info, using time() for uniqueness
my $branch="z9hG4bK" . time();

my $packet = qq(OPTIONS sip:$dst_ip SIP/2.0
Via: SIP/2.0/UDP $my_ip:$my_port;branch=$branch
From: <sip:ping\@$my_ip>
To: <sip:$host>
Contact: <sip:ping\@$my_ip>
Call-ID: $callid\@$my_ip
CSeq: 102 OPTIONS
User-Agent: sip_ping.pl
Date: $date
Allow: ACK, CANCEL
Content-Length: 0

);

# send the packet
print "Sending: \n\n$packet\n" if $verbose;
send($sock, $packet, 0, $portaddr) == length($packet)
    or die "cannot send to $host: $!";
my $send_time = [gettimeofday()]; # start the stopwatch
my $elapsed;

# get the response
eval
{
    local $SIG{ALRM} = sub { die "alarm time out" };
    alarm $RECV_TIMEOUT;
    $portaddr = recv($sock, $packet, 1500, 0) or die "couldn't receive: $!";
    $elapsed = tv_interval($send_time); # stop the stopwatch
    alarm 0;
    1;
} or die($@);

# print our output
if ($verbose) {
    printf("After (\%0.2f ms), host said: \n\n\%s\n", $elapsed*1000,
$packet);
}
elsif ($time) {
    printf("%0.2f\n", $elapsed*1000);
}
else {
    print("$host is alive\n");
}
```

Running the Code

Execute this script using a command like this:

```
# ./sip_ping.pl 192.168.0.123
```

If you see output like this, the SIP host is indeed running:

```
192.168.0.123 is alive
```

Using the -v option with this command gives more verbose output—specifically, the contents of the SIP response and the round-trip latency (in milliseconds) of the SIP message exchange.

You can work this technique into other monitoring tools and run it periodically to inform yourself of any outages. Because the majority of commercial VoIP service providers use the SIP protocol, it can be quite useful. You can even use this script to monitor remote SIP handsets and softphone applications, because all SIP hosts, if implemented correctly, respond in the same manner to the request this script sends.

—Brian Degenhardt

HACK #73 Inspect the SIP Message Structure

SIP is the predominant signaling standard among VoIP carriers and VoIP-enabled PBX systems, so it might be a good idea to know something about it, beyond what the acronym stands for.

SIP is a conversational, connectionless signaling protocol. In English, that means that SIP uses a two-way data conversation, generally using a UDP socket. Its message structure is similar to that of Simple Mail Transfer Protocol (SMTP) or HTTP messages, which also contain headers and a payload. SIP serves many purposes in a telephony environment, including setting up and tearing down VoIP phone calls.

Poking around the VoIP network with Perl is a great way to learn about SIP's message structure. Besides monitoring the availability of hosts, you can use the script from "Monitor VoIP Devices" [Hack #72] as an investigation tool for understanding how the SIP protocol works. Using the -v switch, you can see the full output of a SIP interaction:

```
# ./sip_ping.pl -v 192.168.0.123
```

The preceding command sends the following SIP message to the specified host:

```
OPTIONS sip:192.168.0.123 SIP/2.0
Via: SIP/2.0/UDP 127.0.0.1:6655;branch=z9hG4bK1116720069
From: <sip:ping@127.0.0.1>
To: <sip:192.168.0.123>
Contact: <sip:ping@127.0.0.1>
Call-ID: 0436a2258bedd74d8618e587446810c9@127.0.0.1
CSeq: 102 OPTIONS
User-Agent: sip_ping.pl
Date: Sat, 21 May 2005 05:01:09 PDT
```

```
Allow: ACK
Content-Length: 0
```

The response from the remote host is as follows:

```
SIP/2.0 200 OK
Via: SIP/2.0/UDP 127.0.0.1:6655;branch=z9hG4bK1116720069;received=192.168.0.52
From: <sip:ping@127.0.0.1>
To: <sip:192.168.0.123>
Call-ID: 0436a2258bedd74d8618e587446810c9@127.0.0.1
CSeq: 102 OPTIONS
Contact: <sip:102@192.168.0.123:5060;line=j522azny>
User-Agent: snom200-3.56m
Accept-Language: en
Accept: application/sdp
Allow: INVITE, ACK, CANCEL, BYE, REFER, OPTIONS, NOTIFY, SUBSCRIBE, PRACK,
MESSAGE, INFO
Allow-Events: talk, hold, refer
Supported: timer, 100rel, replaces
Content-Length: 0
```

Notice the plain-text structure of a SIP message packet. Both the request and the response contain very human-readable headers. This is no accident. The body responsible for SIP, the IETF, has a history of advocating human-readable protocols. As a result, SIP avoids machine-friendly ASN.1 encodings such as those used by SIP's predecessors.

The response captured here is from a Snom 200 SIP hardphone. The User-Agent field indicates that it's running version 3.56m of Snom's firmware. Note the Allow, Allow-Events, and Supported headers showing all of the different SIP functionality that this host supports. This is the intended purpose of the OPTIONS request in SIP: to determine the functionality of a remote SIP service.

OPTIONS is just one of several *methods* (by the way, *methods* is SIP's word for requests) that SIP implements. The INVITE method is used to establish calls, and the SUBSCRIBE method is used with presence capabilities like user availability and location. These methods are akin to GET and POST in HTTP. And like HTTP, the host receiving the methods has a variety of numeric responses. In this case, 200 represents that the method was successful, and the response message contains the desired information. Like HTTP, the response for "resource not found" is 404. So, if a SIP INVITE was sent for a user that didn't exist on a particular SIP device, the response would be a 404.

If you point this script at other SIP hosts, you'll see a large variety of responses due to the variation in behavior of different SIP implementations on the Internet and due to the variations in available functionality from one

SIP host to the next. A SIP video-conferencing device might not support the same methods (as listed on the Allow line) as a SIP phone, for example.

See Also

- The official SIP specification at *http://www.faqs.org/rfcs/rfc3261.html*
- *Switching to VoIP* (O'Reilly)
- *Practical VoIP Using VOCAL* (O'Reilly)

—Brian Degenhardt

Audit a Network's QoS Capabilities

#74

Networks without Quality of Service (QoS) measures aren't always suitable to carry voice traffic. So how do you know whether a network path supports QoS?

Using *pathping* and *traceroute* during peak traffic periods, you'll be able to establish whether a particular IP route is a good place for time-sensitive traffic like VoIP media streams. You'll know the jitter and latency qualities of the network, you'll have identified problem routers and potential traffic bottle-necks, you'll know whether each router supports *Resource Reservation Protocol* (RSVP, a QoS standard that allows network bandwidth to be reserved for each call), and you'll know how well the network supports 802.1p prece-dence tagging. (I explain what 802.1p is in this hack; keep going!)

Though Linux is better equipped to provide VoIP services and to serve as a base for troubleshooting, Windows does have a nifty command-line tool that you can use to determine if IP routing supports basic class-of-service measures. *pathping*, which ships with Windows 2000 and Windows XP, lets you see how well your Internet provider—or your corporate network—sup-ports 802.1p and RSVP. This makes *pathping* a particularly useful Windows-only VoIP networking tool.

On non-Windows boxes, though, you still have *traceroute*, of course. While not implicitly a QoS measurement tool, *traceroute* can gather useful perfor-mance data from the VoIP network.

Using pathping

pathping is similar to *traceroute*. It first determines the IP route along all hops from the host where it's running to the host at which it's targeted. Then, it collects information from each hop along the way, like latency times, and displays what information it has collected. The following command returns the hostname and IP address from each hop along the route to the destination, if each hop provides an ICMP response:

```
C:\> pathping www.broadvoxdirect.com
```

The output shows the route to the destination, similar to *traceroute*:

```
Tracing route to www.bigvoxdirect.com [65.67.129.23]
over a maximum of 30 hops:
   0  kelly-6aizy9qd1.ce1.client2.attbi.com [10.1.1.202]
   1  10.1.1.1
   2  10.248.164.1
   3  bic01.elyehe1.oh.attbb.net [24.131.64.38]
   4  12.244.65.61
   5  12.125.176.121
   6  gbr2-p70.phlpa.ip.att.net [12.123.137.26]
   7  tbr1-p012601.phlpa.ip.att.net [12.122.12.101]
   8  tbr1-cl8.n54ny.ip.att.net [12.122.2.17]
   9  ggr2-p300.n54ny.ip.att.net [12.123.3.58]
  10  so-1-0-0.gar4.NewYork1.Level3.net [4.68.127.5]
  11  ge-2-1-0.bbr1.NewYork1.Level3.net [64.159.4.145]
  12  so-0-0-0.mpls1.Cleveland1.Level3.net [209.247.11.134]
  13  ge-6-0.hsa1.Cleveland1.Level3.net [209.244.22.98]
  14  BIGVOX-DIS.hsa1.Level3.net [64.156.66.10]
  15  penguin.bigvoxdirect.com [65.67.129.23]
```

The following example, which uses the -T option, checks to see whether each router along the path supports 802.1p precedence tags. These tags are used by routers and switches to prioritize real-time, delay-sensitive packets such as the UDP datagrams commonly used in VoIP. The more hops along a route that support 802.1p, the better off your VoIP quality on that route is likely to be, because those routers can prioritize voice data over nonvoice data.

```
C:\> pathping www.bigvoxdirect.com -T

Checking for connectivity with Layer-2 tags.

   1  10.1.1.1         OK.
   2  10.248.164.1     OK.
   3  24.131.64.38     OK.
   4  12.244.65.61     OK.
   5  12.125.176.121   OK.
   6  12.123.137.26    General failure.
```

The output from *pathping* first shows the route, like the previous example, but also adds the 802.1p feedback as far along the route as possible. Not all devices along every route support 802.1p. In this example, the sixth hop does not, because the router isn't configured for IP Type of Service (ToS) and 802.1p. Since the 802.1p header can't be carried past the sixth hop, subsequent hops cannot be tested for 802.1p support.

The -R option will do a similar check for RSVP support, in a similar fashion. You aren't nearly as likely to find RSVP-supporting hops on the public Internet as you are 802.1-aware hops. But, if RSVP is configured on a private network, you can use *pathping* to help you evaluate that network's hardware readiness for QoS. It will tell you which routers support RSVP and which routers need to be either reprogrammed or upgraded to support it.

Measure the Latency Time and Jitter on a Call Path

The cumulative latency on a route is a good indicator of how latent it is and, therefore, how well it will work as a VoIP call path. An easy way to record latency between hops (routers) on a route is by using *traceroute* (on Windows, *tracert*).

Using *traceroute*, you can discover the route to the host at the specified address, send several ICMP packets to each hop on the route, and then be shown the following:

- The highest round-trip latency to each router, in milliseconds
- The lowest round-trip latency to each router
- The average round-trip latency to each router
- The IP address and/or hostname of each router
- Whether an ICMP ping response was received from each router

The syntax for *traceroute* is very simple:

```
# traceroute www.macvoip.com
Tracing route to www.macvoip.com [65.31.69.11]
over a maximum of 30 hops:

   1     1 ms      1 ms      1 ms   10.1.1.1
   2    14 ms     13 ms     18 ms   10.248.164.1
   3    18 ms     16 ms     12 ms   bic01.elyehe1.oh.attbb.net [24.131.64.38]
   4    19 ms     21 ms     34 ms   12.244.65.61
   5    31 ms     23 ms     24 ms   12.244.72.70
   6    25 ms     26 ms     28 ms   tbr1-p012401.phlpa.ip.att.net [12.123.137.45]
   7    32 ms     27 ms     27 ms   tbr1-cl8.n54ny.ip.att.net [12.122.2.17]
   8    28 ms     28 ms     34 ms   ggr2-p300.n54ny.ip.att.net [12.123.3.58]
   9    31 ms     30 ms     28 ms   att-gw.ny.aol.net [192.205.32.218]
  10    32 ms     28 ms     43 ms   bb2-nye-P1-0.atdn.net [66.185.151.66]
  11    29 ms     47 ms     34 ms   bb2-vie-P12-0.atdn.net [66.185.152.201]
  12    64 ms     48 ms     62 ms   bb2-chi-P6-0.atdn.net [66.185.152.214]
  13    60 ms     60 ms     62 ms   RR-DET.atdn.net [66.185.141.98]
  14    59 ms     54 ms     66 ms   os0-0.fmhlmi1-rtr1.twmi.rr.com [24.169.225.65]
  15    57 ms     53 ms     63 ms   ig0-1.fmhlmi1-ubr5.twmi.rr.com [24.169.225.22]
  16    64 ms     66 ms     68 ms   www.thelinuxfix.com [65.31.69.11]
```

Whether on Linux, Windows, or Mac, *traceroute*'s output tends to be the same. This sample output is from Windows, but all *traceroute*s show you the minimum, average, and maximum latency to each hop along the route.

Not all IP networks permit ICMP traffic—or *traceroute*s in particular—because some system operators prohibit them for security reasons. Most routes across the Internet should provide a valid response when using the *traceroute* command. As you examine the output from the *traceroute* command, pay special attention to the variance in highest and lowest latency times (*not* the average latency time). This variance is a good, rough estimate of jitter between each hop. (If you don't know what jitter is, don't freak out! Just refer to "Sniff Out Jittery Calls with Ethereal" [Hack #83].)

Keep in mind, though, that the latency and jitter patterns of VoIP traffic (which is UDP and very consistent in nature) can vary from your readings here, since *traceroute* uses ICMP packets that are very bursty in nature. The routers along the route that you're evaluating might already be configured to treat VoIP traffic with greater precedence, but at least you'll get an idea of the general service conditions along the route.

See Also

- *http://www.microsoft.com/technet* for more information on *pathping*

HACK #75   Graph Latency and Jitter

Use sip_ping.pl to record latency and jitter data in a nice, pretty graph.

It's time to step beyond the basic *ping* and *traceroute*, and to graph packet flow using Perl. In "Monitor VoIP Devices" [Hack #72], you saw how to determine the availability of a SIP host using Perl and the SIP OPTIONS packet. The next logical step is to time how long it takes to receive a response to monitor the latency on the link between you and the device you're monitoring. This hack uses the script from "Monitor VoIP Devices" [Hack #72], but this time you use the -t option to time the round trip:

```
# ./sip_ping.pl -t 192.168.0.123
50.77
```

This shows me that it took almost 51 milliseconds for my Snom handset to respond to a SIP request. This isn't a terribly useful test, however, considering I'm timing this response from my desktop computer hooked up to phone's built-in switch. Latency (and its cousin, jitter) starts to be an issue when your VoIP traffic passes out of your local area network (LAN) and traverses the wide Internet. For every router that your VoIP traffic passes through, there is a chance for your VoIP packets to be delayed or, worse, dropped. You can use the *traceroute* (*tracert* on Windows) application to see how many hops it takes to get from your computer to a remote VoIP provider. Here is an example from my PC to a SIP-based Internet telephony provider:

```
# traceroute sip.example.com
traceroute to sip.example.com (172.16.15.15), 30 hops max, 38 byte packets
 1  192.168.0.1 (192.168.0.1)  0.213 ms   0.189 ms   0.194 ms
 2  10.69.128.1 (10.69.128.1)  12.113 ms  12.343 ms  22.133 ms
 4  10.25.25.25 (10.25.25.25)  7.280 ms   7.777 ms   9.138 ms
 5  sip.example.com (172.16.15.15)  74.124 ms  77.450 ms  77.033 ms
```

This shows that my VoIP traffic has to pass through five routers every time I make a call. If any router gets congested or starts having problems, my call quality can suffer. Though knowing about troublesome routers along the path is important, to keep things simple, you can focus most of your concern on the time delay between yourself and the destination host. To determine this you can use the *ping* utility:

```
# ping sip.example.com
PING sip.example.com (172.16.15.15) 56(84) bytes of data.
64 bytes from 172.16.15.15: icmp_seq=0 ttl=53 time=74.4 ms
64 bytes from 172.16.15.15: icmp_seq=1 ttl=53 time=74.4 ms
64 bytes from 172.16.15.15: icmp_seq=2 ttl=53 time=79.0 ms
64 bytes from 172.16.15.15: icmp_seq=3 ttl=53 time=75.4 ms
64 bytes from 172.16.15.15: icmp_seq=4 ttl=53 time=77.8 ms
```

This shows that there is a 74 to 78 ms delay between my computer and the remote VoIP server. This delay is called *latency*. In the context of VoIP traffic, it's not always a bad thing to have consistent latency. Imagine if I consistently have 100 ms of latency on my call. That is, it takes one-tenth of a second for my speech to reach the ears of the person I am talking to. This isn't terribly noticeable. However, if this 100 ms delay suddenly evaporated to 50 ms and then jumped back to 100 ms, this would definitely create audible abnormalities in my speech. Every time it sped up, the audio would skip over the slower packets that hadn't arrived yet. Every time it slowed down, there would be a slight pause, waiting for more audio to arrive. This is called *jitter*, and this is the true source of quality problems on the VoIP frontier. However, this is not to say that latency is not an important measurement. Latency measurements can be a basis for measuring the potential for jitter. For example, say I have two hosts: one with 200 ms of latency and one with 5 ms of latency. A 20% variation in latency will result in 40 ms of jitter from the first host, and only 1 ms of jitter from the second.

One way to deal with jitter is to use a *jitter buffer*, a device that basically delays playing or sending on the audio packets for a short period of time to cushion against any delayed or out-of-order packets. As the packets arrive, the buffer grows and shrinks to accommodate variations in their latency, thus smoothing out their perceived latency. While jitter buffers are an excellent tool for improving audio quality on VoIP traffic, they still come with the cost of added latency in the call. It is preferable to eliminate the jitter on the network altogether, if possible.

One big source of jitter and latency is network congestion. Let's suppose that you're on an important VoIP call, when somebody on your network decides to download the latest movie trailer, instantly using up all of your bandwidth. Suddenly your VoIP packets have to wait in line behind the movie trailer packets, causing a change in latency—i.e., jitter. One solution to this problem is to use QoS policies on your router. This is the practice of prioritizing some types of network traffic ahead of others. QoS works well because some network traffic, like downloading movie trailers or checking your email, is not affected by small changes in latency or jitter, and other services, like VoIP traffic, are.

This uncovers a flaw in our use of *traceroute* and *ping* to measure latency. Because ping traffic is not the same as VoIP traffic, some routers' QoS policies might treat them differently. Hence the need for our Perl SIP *ping* utility [Hack #72]. As it uses SIP, it provides a much better estimate for measuring VoIP latency and jitter. In addition to network latency, it will also measure any latency injected into the system by the SIP application listening on the other end.

However, since SIP-based VoIP relies heavily upon another protocol altogether to carry the actual audio streams—Real-time Transport Protocol (RTP)—measuring latency and jitter merely by sending SIP messages isn't foolproof. Yet, since SIP and RTP are almost always UDP (and ping packets are not), this kind of measurement is better than using *ping* and *traceroute*. Plus, many VoIP-aware routers give the same preference to SIP as they do to RTP, so the results you get using this technique might not be too far off.

The Hack

Timing the round trip of SIP with Perl is only half of this hack. The second half is using a Unix program called *RRDtool* (*http://people.ee.ethz.ch/~oetiker/webtools/rrdtool/*) to graph the data. *RRDtool* is a generic utility for graphing data over time. You'll need to build it by following the instructions at the author's web site.

Once you've installed *RRDtool* on your Linux PC, you'll use it to graph the SIP latency of three different SIP providers. The first thing to do is create the *rrd* database file, which *RRDtool* will use to accumulate the data you'll graph later:

```
$ rrdtool create voiphacks.rrd -s 300 DS:provider1:GAUGE:300:U:U \
DS:provider2:GAUGE:300:U:U DS:provider3:GAUGE:300:U:U \
RRA:MAX:0.5:1:10000
```

This says that we will take 10,000 samples of SIP ping time. One ping will happen every 5 minutes until 10,000 pings have occurred. There are three data sources, one for each of three providers that will be guinea pigs with which to build the graph. You will notice that this creates a file called *voiphacks.rrd*. This is the database file that will be storing the latency measurements. The next step is to make the measurements. The following shell script will launch an instance of *RRDtool*, taking latency measurements from the three providers using our *sip_ping.pl* utility once every 5 minutes:

```
#!/bin/sh
SIP_PING="../sip_ping.pl -t"

while /bin/true; do
    PROVIDER1=`${SIP_PING} sip.provider1.com 2>/dev/null || echo "INF"`
    PROVIDER2=`${SIP_PING} sip.provider2.com 2>/dev/null || echo "INF"`
    PROVIDER3=`${SIP_PING} sip.provider3.com 2>/dev/null || echo "INF"`
    rrdtool update voiphacks.rrd N:${PROVIDER1}:${PROVIDER2}:${PROVIDER3}
done
```

You can run this in the background on a Unix system. Don't forget to make your script executable (**chmod 755 *filename.sh***) and put it in a place where it can be seen by any startup scripts if you decide to have it run on system boot. If you'd like to run the script periodically, you can add it to your */etc/crontab* file so that the cron daemon can run it automatically according to your own timetable. After you have accumulated a day's worth of data, you can graph the results:

```
$ rrdtool graph pingtimes_daily.gif -v "milliseconds" \
--title="Latency over 1 day" -s now-1d -w 875 -h 475 \
DEF:provider1=voiphacks.rrd:provider1:MAX \
DEF:provider2=voiphacks.rrd:provider2:MAX \
DEF:provider3=voiphacks.rrd:provider3:MAX \
LINE2:provider1#882222:"Provider 1" LINE2:provider2#228822:"Provider 2" \
LINE2:provider3#222288:"Provider 3"
```

This will produce a graph of a full day's results, taken at five-minute intervals (Figure 6-1). If you'd like to graph a longer period of time, you can change the -s option to something like now-1w for a week, or now-1m for the results graphed over a month. The manpage for *rrdgraph* has some more examples that show how to tweak the output.

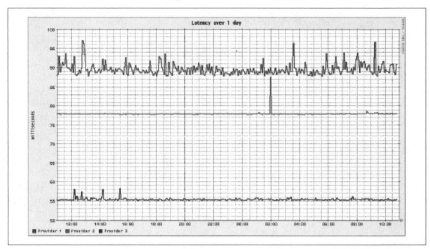

Figure 6-1. A graph of latency and jitter across a VoIP link

This graphs latency measured over time. To determine jitter, you can examine the variation in the line on the graph for each host. If it's a consistent 80
ms flat line, there's little likelihood of jitter. If it bounces all over the place,
it's more likely that you'll experience jitter on a call.

Because this script takes a sample only every five minutes, it doesn't provide
the truest possible measurement of the type of jitter that will affect a VoIP
call. But since it's very difficult to trigger UDP datagrams at a real-world rate
of 20 to 50 per second using the *sip_ping.pl* and *RRDtool* tools, measuring
latency over time will have to suffice. Plus, with that many samples, you'd
have a ton of numbers to crunch to figure out the utilization trends over a
long span of time, like days, weeks, or years. By compromising short-term
accuracy and taking samples only every five minutes, it's easy to look at
latency over long spans of time.

—Brian Degenhardt

HACK #76 Explore NAT Traversal

Network Address Translation (NAT) poses a fundamental problem for VoIP.
But like so many problems, the first step toward a solution is realizing you've
got a problem.

Have you ever tried a Voice over IP application that purports to let you talk
to your buddies over the Net, only to fire it up and not be able to call them
or, worse yet, to hear only dead silence on the other end? Chances are, the
VoIP application is being broken by your firewall router.

NAT is most often used to permit computers that are on private networks behind firewall routers to access services on the public Internet without having to have public IP addresses.

The Internet is designed such that most private networks that have access to the Internet do so using a device called a NAT firewall. By using NAT, computers on the private networks are afforded a certain amount of security when accessing the Internet, because the NAT firewall can block certain network services and log all access attempts. The most common form of NAT is called *masquerading*, which uses a public IP address on the firewall's external interface to conceal all of the computers connected to the private interface. In this way, those computers "masquerade" as the firewall to the outside world.

The problem with this practice is that it breaks applications that rely on that public IP address. When a private computer makes a request, the server handling the request might attempt to respond to the requester using the firewall's public IP address rather than the requestor's private IP address, which is unknown to the responding server. This wreaks havoc on many protocols, including the file transfer protocol (FTP) and SIP.

Indeed, SIP's poor NAT traversal capabilities are famous. That is, if you have a router that does NAT, as most broadband routers do, it can potentially cause problems with using SIP. This is because the SIP protocol requires SIP hosts to instruct each other on how they can be reached by using the Via header, which looks like this:

```
Via: SIP/2.0/UDP 127.0.0.1:6655
```

This can cause issues when your SIP host places its internal IP address in the Via header instead of the external address of the NATing router. There are a plethora of solutions to this problem, including Simple Traversal of UDP NATs (STUN) and RFC 3581. STUN is a client-side technique used to determine the correct external IP address and port, which are then placed in the Contact header. RFC 3581 is a server-side solution where the server responds to the host that sent it the packet instead of the host in the Contact header. You can see this in action by using the *sip_ping.pl* script from the previous hack:

```
# ./sip_ping.pl -v proxy01.sipphone.com | grep Via
Via: SIP/2.0/UDP 127.0.0.1:6655;branch=z9hG4bK1116737044
Via: SIP/2.0/UDP
127.0.0.1:6655;rport=14328;received=66.27.57.228;branch=z9hG4bK1116737044
```

Note the addition of the rport and the received parameters in the response of this ping. These show the remote port and IP address, respectively. While our client stated to the server that its Contact is 127.0.0.1 port 6655, the remote host decided to send the packet back the way it came, to 66.27.57.228 on port 14328.

Some router manufacturers have tried to get into the game and transparently rewrite SIP packets that pass through them. Although this was borne out of good intentions, it can cause problems on certain SIP implementations when the returning response does not match what was transmitted. If you're trying to use SIP behind a firewall and you aren't aware that that well-meaning firewall has altered your transmissions, it will be next to impossible to get around the fundamental problem of SIP using other, more generally accepted solutions (like STUN; more on that later). Here is an example of a SIP packet sent through a router with SIP translation enabled:

```
# ./sip_ping.pl -v proxy01.sipphone.com | grep Via
Via: SIP/2.0/UDP 127.0.0.1:6655;branch=z9hG4bK1116737081
Via: SIP/2.0/UDP 10.1.1.5:6655;branch=z9hG4bK1116737081
```

Note that since the packet was rewritten, the remote host's RFC 3581 implementation did not insert the received and rport fields. However, when the packet returned, a different IP address was inserted in the Via header. This can be especially troublesome on hosts with multiple IP addresses, as the router rewriting the headers might be assuming that all responses should go to one host. This can be a good thing when you have a single SIP device on the private network, but it's a bad thing when you've got a whole building full of SIP phones. If the router is rewriting SIP headers arbitrarily, it won't be possible for all of these phones to receive their SIP responses.

The *sip_ping.pl* script takes two additional arguments (-s, for source, and -p, for port) that you can use to explore NAT traversal. These are the source IP address and the source port to be placed in the Via and Contact headers. By taking the rport and received addresses of the preceding example, we can rewrite the headers correctly to make it appear as though our packet is coming from the external interface of the NAT firewall:

```
# ./sip_ping.pl -v -s 66.27.57.228 -p 14328 proxy01.sipphone.com | grep Via
Via: SIP/2.0/UDP 66.27.57.228:14328;branch=z9hG4bK1116737949
Via: SIP/2.0/UDP 66.27.57.228:14328;branch=z9hG4bK1116737949
```

This now specifies the correct external IP address and port in the Via header. By specifying the public IP and port, we can now traverse NAT, even when contacting remote hosts that do not support RFC 3581. This is a great example of how we can get around the NAT problem with SIP.

Now, if only our SIP client—the phone—were smart enough to employ this technique itself, our SIP phone would appear as though it were outside the firewall to the remote SIP devices it calls. That's where the STUN protocol comes into play.

Get STUNned

STUN is an IETF recommendation that allows SIP clients to connect to a STUN server on the Internet to determine what address and port the connecting client appears to have from behind a NAT host. This way, the client can send the appropriate SIP Via headers, and, at least in theory, the SIP devices it calls will be able to respond by sending packets to the client's NAT firewall. Once a SIP device sends response packets to the NAT firewall, it's up to the firewall to forward them to the appropriate client on the private network.

So, there are two steps to hacking out a functional STUN server. First, set up and configure the server. Second, configure your NAT firewall so that it forwards inbound VoIP traffic to your SIP phone.

To build a STUN server, you'll need a Unix box (Linux, BSD, Mac OS X, etc.) that has a non-NATted connection to the Internet—i.e., it has a public IP address. There is a Win32 version of STUN, too, but compiling on Unix is a lot less trouble. You can obtain the STUN server software from *http://sourceforge.net/projects/stun/*. Download and unpack it into your Unix machine's */usr/src/* directory; then compile and install it like this:

```
# wget http://kent.dl.sourceforge.net/sourceforge/stun/stund_0.94_Oct29.tgz
# cd /usr/src
# tar xvzf stund_0.94_Oct29.tgz
# cd stund
# make
# cp server /usr/sbin/stund
# /usr/sbin/stund &
```

That last command launches the STUN server. Make a note of your STUN server's internal IP address (on a small network, this is probably your default gateway), as you'll need to provide it to your SIP phone as the STUN server address.

But What About RTP?

So, STUN will allow your SIP client to know what to put in the SIP Via header, but remember that the VoIP call also relies upon RTP (an entirely different protocol) for transporting the voice packets. These packets must

flow on two separate UDP sockets: one from you to the person you're calling and one from that person back to you. NAT breaks this, too! To make
sure your SIP phone can receive voice packets behind the NAT box, you'll
need to tell the NAT box to forward incoming RTP traffic to your SIP
phone's IP address.

For the sake of this example, I'm going to use the X-Lite softphone, a free
STUN-aware SIP softphone for Mac and Windows. Download it from *http://
www.counterpath.com/* and install it. (Review "Use a Softphone with a VoIP
TSP" [Hack #4] for a quick refresher.) If you want to attempt this hack with
another softphone, be sure it supports both SIP and STUN, or no cigar.

Get the IP address of the computer where X-Lite is installed. Make a note of
it (I'll use 10.1.1.50 for this example). You'll need to tell your NAT firewall
to forward all VoIP traffic to this address. If your NAT box runs Linux, you
can use the following commands to forward all the inbound VoIP traffic to
10.1.1.50—your IP phone. This assumes the Internet-facing interface is eth0
and its address is 201.101.1.1, so replace that address to suit your firewall:

```
# iptables -t nat -A PREROUTING -p udp -i eth0 -d 201.101.1.1 \
       --dport 5060 -j DNAT --to 10.1.1.50:5060
# iptables -A FORWARD -p udp -i eth0 -d 10.1.1.50 --dport 5060 \
       -j ACCEPT
# iptables -t nat -A PREROUTING -p udp -i eth0 -d 201.101.1.1 \
       --dport 5061 -j DNAT --to 10.1.1.50:5061
# iptables -A FORWARD -p udp -i eth0 -d 10.1.1.50 --dport 5061 \
       -j ACCEPT
# iptables -t nat -A PREROUTING -p tcp -i eth0 -d 201.101.1.1 \
       --dport 5000 -j DNAT --to 10.1.1.50:5000
# iptables -A FORWARD -p tcp -i eth0 -d 10.1.1.50 --dport 5000 \
       -j ACCEPT
# iptables -t nat -A PREROUTING -p tcp -i eth0 -d 201.101.1.1 \
       --dport 5001 -j DNAT --to 10.1.1.50:5001
# iptables -A FORWARD -p tcp -i eth0 -d 10.1.1.50 --dport 5001 \
       -j ACCEPT
```

Now, any inbound SIP (ports 5060 and 5061) or RTP (ports 5000 and 5001
in this example) traffic will be forwarded directly to host 10.1.1.50, where
your SIP softphone is running. (Ordinarily, you wouldn't need to forward
the SIP ports for outbound calls to work; STUN takes care of that as long as
the SIP client can use STUN.)

For a list of VoIP-related port numbers, refer to "Log VoIP
Traffic" [Hack #84].

Next up is configuration of the STUN client in the softphone. Launch X-Lite
and configure it as you normally would; then add the private address of your

NAT machine, which is now running a STUN server as well, to the Primary STUN Server entry of X-Lite's preferences dialog. Now, call somebody outside your private network to see if it works (if that person is also behind a NAT, it might not!).

Shape Network Traffic to Improve Quality of Service

There's a reason why the Public Switched Telephone Network (PSTN) has been the dominant global voice network for a century: Quality of Service.

VoIP is a wonderful technology. It enables all kinds of features and portability options that are not available with traditional telephony technologies. However, unlike traditional telephony, VoIP has some inherent quality issues. By the time you finish reading this hack, you'll know how to tackle the quality problem with the VoIP engineer's best weapon: the Linux kernel's built-in router.

On the traditional telephone network, every single call has a dedicated time slot using a technology called *time division multiplexing* (TDM). With TDM, a circuit is divided into several time slots, each with its own dedicated slice of bandwidth. This is what ensures that your call is the only call in that time slot, and that after all of the time slots are used, the circuit is at top capacity and no further calls will be allowed.

With VoIP, your call is converted into thousands of small datagrams (or packets, if you please). These packets are then queued up on a device (your computer, analog telephone adapter [ATA], router, etc.) and thrown out over the wire, with no guarantee that they will even reach their ultimate destination, wherever that might be. You can see how this might cause problems with voice quality, especially when other data traffic on that same link is vying for that link's limited capacity (or "bandwidth," depending on your preferred vernacular).

This is especially problematic with consumer access technologies such as cable modem or DSL. In a typical residential or small-office setup, you will have one relatively high-speed link to the Internet, and that link is responsible for carrying email, web surfing, and even the occasional big download of a CD-ROM image or something. Now try to put voice traffic on this link.

Humans are very sensitive to delay when listening to speech. If one web site on your computer loads 250 ms slower than another web site, you're not really going to notice. However, if there is a 250ms delay in a conversation, you'll perceive that as a very annoying delay, and it will make ordinary conversation difficult.

With all of this traffic on one link, how can I make sure that someone down-loading a song from iTunes does not cause the audio on my VoIP call to suf-fer? It's easy, with a technology referred to as Quality of Service (QoS). QoS is a general term applied to a family of technologies that essentially manipulate the first in, first out (FIFO) queues on devices. Remember that PC from before, or that router or ATA? Normally, all of the IP traffic from that device will be placed into a FIFO queue for delivery to the remote endpoint. With QoS, we can manipulate that queue and pass judgment on packets match-ing certain attributes that move them to the front of the line, regardless of what time they got in, because they are more important to us. This is what you'll do with your VoIP packets.

A router with Linux (i.e., a PC with two Ethernet interfaces and Linux installed) will help us. When you place this router between two networks—such as the Internet and your LAN—you can use it to enforce QoS measures.

Thanks to the wonderful folks at *http://www.lartc.org/*, I was able to create a traffic-shaping script that works very well for prioritizing VoIP traffic. It is called AstShape, and it is included in the AstLinux distribution. However, you can use it in any Linux distribution that includes *iproute2*. This should be just about any major, modern distribution out there today.

The first thing you need to do is visit my web site, *http://www.kriscompanies. com/*, and click on Downloads → Asterisk → AstShape. The AstShape script will begin downloading, and should finish very quickly. After you have saved it to your hard disk, copy it onto your router in a place like */usr/local/ sbin*. Whatever directory you pick, make sure that it is in your $PATH. You can see your $PATH by executing **echo $PATH** from the command prompt.

Once you have AstShape "installed," make it executable by running the following:

```
# chown root:root  astshape ; chmod 750 astshape
```

This will ensure that no one other than root can run this script. Open Ast-Shape in your favorite text editor, and take a look around. I will show the first few lines of AstShape and explain their meanings:

Set this to the downstream speed of your connection (in kilobits). Use a speed test like the one available at *http://toast.net/* to get an accurate idea of your connection's actual speed; for instance, the overhead of PPPoE accounts for an approximate 10%–13% drop in speed from what is adver-tised by many consumer DSL packages. Test and test often! Also, you will want to set this number to about 85% of your actual test speed. (See the sidebar "VoIP QoS" for an explanation.)

```
DOWNLINK=5500
```

VoIP QoS

Most broadband service providers configure their networks for bulk traffic speed. They know that to most customers, speed is measured by how many KB/s their web browser displays when downloading a large file. However, this is not the whole story. With VoIP, a measurement called *latency* is far more important.

The best possible way that I have ever heard to describe the concept of latency is the Concorde (R.I.P.) versus Boeing 777 analogy. The British Airways Concorde can get 92 people from New York to London in about 3.5 hours. The Boeing 777 can get 440 people from New York to London in about 6.5 hours. Which is faster? If you had to transfer a large number of people (using only one plane), the 777 would be "faster," even though it travels at half the air speed of the Concorde. If you had to transfer a small number of people very quickly, the Concorde would be "faster." This applies to VoIP very well. When you are downloading a large file from a remote web server, you will be dealing with fewer, very large packets. With VoIP, you are dealing with many more, much smaller packets. In fact, with some VoIP codecs, the size of the Ethernet/IP/UDP header is much larger than the codec payload itself (G729 being a good example)!

So, VoIP is the Concorde and most everything else is the Boeing 777. What does this have to do with limiting the speed of your connection by 15%? Simple. By limiting the speed of your connection by 15%, we are (hopefully) ensuring that the FIFO queues outside of your control do not fill up completely. Anyone who has ever used VoIP on a cable modem or DSL line knows what happens when someone else using that connection begins downloading a very large file. The user on the VoIP connection experiences large gaps in audio transmission, sometimes lasting several seconds. This is because the FIFO queues on your cable/DSL modem—and on your Internet Service Provider's (ISP's) CMTS/DSLAM—fill completely with web traffic, and your tiny little VoIP packet is at the end of the line. Because we can't control these FIFOs like we can our Ethernet interface, we have to place a hard limit on the amount of traffic.

However, not all hope is lost. If you are an ADSL subscriber using Linux, you should look into the S518 ADSL board from Sangoma Technologies. For around $115 USD, you can have an internal PCI form factor ADSL modem that you have *complete* control over. When you use it with the PPPoE client software from Roaring Penguin, you can eliminate your SpeedStream kludge of a modem and gain the enhanced speed, logging, and feature set provided by the S518. I highly recommend it to anyone already using ADSL and Linux. Plus, you don't have to cap your link speed at 85%, as the queuing on the S518 can be controlled from Linux!

Set this to the upstream speed of your connection. Use the Internet speed test results from before, and again subtract 15% from your results. The best way to determine this number is by testing, testing, testing:

```
UPLINK=550
```

This is your external network device—probably eth0 or eth1. However, if you are using the Sangoma S518 mentioned in the sidebar "VoIP QoS" (or dial-up—in which case you must really love torturing yourself), this will probably be ppp0:

```
DEV=eth1
```

This is a list of ports, separated by spaces to be added to the VoIP class. This class of traffic is given highest possible priority. 4569 is the port for IAX2, Asterisk's native Inter-Asterisk Exchange (IAX) protocol. Do not—*do not*—put 5060 here, ever (more on this later)!

```
VOIPPORTS="4569"
```

This is a list of ports to be given "interactive" priority. This is the next highest level of priority, and by default it includes two common ports used for SIP signaling. Please note that with common SIP devices, signaling (SIP) and audio transmission (RTP) take place in two (or more) separate UDP connections and do not travel on the same port. Many people make the mistake of adding port 5060 to their highest class of service for QoS. This does nothing for audio quality, and merely assures that SIP messaging (call setup, status, etc.) is given highest priority. While SIP is a time-sensitive protocol, RTP audio is much more so! Also, you might be thinking of adding port 22 (SSH) to this list. Don't do it just yet. You'll need to have more tricks up your sleeve for SSH:

```
INTPORTS="5060 5061"
```

This is a list of source ports to be added to the "bulk" class of service. This should include all traffic that tends to be large, sustained downloads/uploads. You might ask why port 22 (SSH) is listed here. As I mentioned before, we have some special instructions for SSH later on. Adding port 22 here essentially covers file transfers using SSH, not SSH shell sessions:

```
NOPRIOPORTSRC="25 22 80 110 143 943"
```

This is the same as the NOPRIOPORTSRC line, except it refers to destination ports:

```
NOPRIOPORTDST="25 22 80 110 143 943"
```

The Actual Script

Here I will go over the actual commands from AstShape and attempt to break them down. If you are not interested in modifying AstShape beyond adjusting the preceding values, you do not need to read this section. If you need to do more tweaking, or are just plain curious, read on!

This line installs the root hierarchical token bucket (HTB) queue and points default traffic to the 30 class:

```
tc qdisc add dev $DEV root handle 1: htb default 30
```

This line defines the queue used for VoIP. As I say in the script, this is the "Crown Prince of Bandwidth." Nothing has higher priority than VoIP in AstShape:

```
tc class add dev $DEV parent 1:1 classid 1:10 htb rate \
${UPLINK}kbit burst 6k prio 1
```

The same for the "interactive class":

```
tc class add dev $DEV parent 1:1 classid 1:20 htb rate \
${UPLINK}kbit burst 6k prio 2
```

The default class:

```
tc class add dev $DEV parent 1:1 classid 1:30 htb rate \
$[9*$UPLINK/10]kbit burst 6k prio 3
```

The bulk class:

```
tc class add dev $DEV parent 1:1 classid 1:40 htb rate \
$[8*$UPLINK/10]kbit burst 6k prio 4
```

Now that we have our queues defined, we need to assign traffic to them.

Any IP packets with TOS=0x18 belong in the VoIP class:

```
tc filter add dev $DEV parent 1:0 protocol ip prio \
10 u32 match ip tos 0x18 0xff flowid 1:10
```

Any IP packets with TOS=0x10 (minimum delay) belong in the interactive class:

```
tc filter add dev $DEV parent 1:0 protocol ip prio \
20 u32 match ip tos 0x10 0xff flowid 1:20
```

By default, most SSH client/server programs will set the IP TOS field to 0x10. How convenient!

Add DNS to "interactive," too:

```
tc filter add dev $DEV parent 1:0 protocol ip prio \
21 u32 match ip sport 53 0xffff flowid 1:20
tc filter add dev $DEV parent 1:0 protocol ip prio \
22 u32 match ip dport 53 0xffff flowid 1:20
```

DNS is a very time-sensitive protocol, where delays in name resolution can usually be noticed by a user or application very easily. Also, DNS queries are not very large, so it is to our benefit to add them to a higher class of service.

You will see several lines that talk about Transmission Control Protocol (TCP) ACKs. These are TCP acknowledgments, and they won't be covered in this book. Trust AstShape (and me) by leaving this alone.

Finally, we assign whatever is left to the earlier default class:

```
tc filter add dev $DEV parent 1: protocol ip prio \
30 u32 match ip dst 0.0.0.0/0 flowid 1:30
```

As you can see, AstShape is a very simple yet powerful traffic QoS script. A big thanks goes to the folks at LARTC for providing WonderShaper, which AstShape was based on. For more information on traffic shaping/QoS under Linux, please visit the LARTC web site at *http://www.lartc.org/*.

—*Kristian Kielhofner*

HACK #78   Create a Premium Class of Service

You can produce high-octane voice service for premium users by building distinct classes of service.

In this hack, we will be using the AstShape script that we played with in "Shape Network Traffic to Improve Quality of Service" [Hack #77]. However, you'll now be prioritizing just one class of *voice* traffic over another. This is particularly useful when you need to segment two groups of users, regardless of the applications they're running. In the other hack, I showed you how to prioritize voice for everybody. But suppose you want to prioritize it only under certain circumstances. If you were going to launch a service like Skype, where users can make free calls to other Skype users and pay for calls to non-Skype users, you would want to provide the highest possible quality for the paying users, right?

Let's say that you have a VoIP service that allows callers to interconnect with the PSTN (like Skype) and with other VoIP users on the Internet. Let's also say that you have two pricing levels. The "economy" pricing level does not guarantee quality (and is less expensive, or perhaps free), but the "premium" pricing level does (and costs more in return). I will show you how you can implement this using a slightly modified version of AstShape.

There are two ways Linux traffic control can build classes of service: by port number and by ToS headers.

First, I am going to assume you've got a PC with two Ethernet interfaces loaded up with Linux, and the ability to control the IP ToS bits *or* port numbers used by the VoIP devices you're going to support, be they IP phones or

VoIP servers. This is absolutely critical. All we are going to use to separate the two levels of our traffic are the IP ToS bits or UDP/TCP port numbers, so without that ability, this hack will be less than useful (and not much fun either).

But how do you know if you can control the port numbers used by VoIP applications? Just about all VoIP software lets you control the port numbers employed by the signaling protocol (SIP) and the voice stream (RTP). Asterisk lets you adjust these settings in */etc/asterisk/sip.conf*, and IP phones and softphones like X-Lite have user-configurable port-number preferences.

> To view or adjust the range of RTP ports used by Asterisk, take a peek inside */etc/asterisk/rtp.conf*.

As I mention in other hacks, SIP is just the call signaling protocol. RTP is the protocol that carries voice and other data (video, images, data, etc). The port numbers used by RTP are pseudo-randomly selected from a predefined port range. In Asterisk, this can be configured in */etc/asterisk/rtp.conf*. Actually, the default range of 10000–20000 is considered by many to be too wide for most installs. Please adjust according to the size of your installation.

Altering the ToS bits can be a little trickier. Most IP phones and VoIP services tag media packets with the highest possible priority, so forcing them to "downgrade" some of the packets into a lower class of service is hard. Ideally, you'll build your premium and economy service classes using the port number rather than the ToS bits.

Once you have met these conditions, you are ready to proceed. Surf over to *http://www.kriscompanies.com/*. Go to Downloads, and then Asterisk, and then locate AstShape (Provider). Download it to your machine, place it somewhere in your $PATH (like */usr/local/sbin*), make it executable (**chmod +x astshape-provider**), and optionally change the name to something that you will remember. Let's take a look at the script, shall we?

Get Started with AstShape Provider

Open AstShape Provider in your favorite text editor. If you have ever seen AstShape, you will notice that AstShape Provider is actually smaller and simpler. That's because we are assuming that all this router will handle is VoIP traffic. There are no provisions for handling other types of traffic, and, as the script says, you will want to block this traffic with *iptables* or some other firewall. There are four possible knobs to turn, and they look conspicuously like those in plain-vanilla AstShape.

This is the speed (in kilobits) of your Internet connection. This value can be best determined by testing, and testing often. This will be the hardest part:

```
LINKSPEED=1000
```

This is the wide area network (WAN) interface on which to do QoS:

```
DEV=eth1
```

What you have here is a list of ports, separated by spaces, that will be placed in "Class 1." This is the premium (more important) class of service. I've chosen 5000 and 5001 for my premium class's ports:

```
#Class 1 priority ports
CLASS1PORTS="5000 5001"
```

Next, you have a list of ports, separated by spaces, that will be placed in "Class 2," the economy class. In this class, I've put ports 10000 and 10001:

```
#Class 2 priority ports
CLASS2PORTS="10000 10001"
```

Once you have these values set to the correct port numbers, save the script and exit. Now all you have to do is run it:

```
# ./astshape-provider
```

You shouldn't see any errors. The `./astshape-provider` status command will list the queues that have been defined.

Explaining the AstShape Provider Script

What is this doing? How does this work? What if I need more? Hold on! Slow down! The AstShape Provider script is actually quite simple. Let me break it down for you on a line-by-line basis:

Here you can see that the first packet queue set up by `tc` is known as the root queue. I've told `tc` to use HTB queuing. This is just one of many packet-queuing techniques supported by the Linux kernel:

```
tc qdisc add dev $DEV root handle 1: htb default 30
```

This says to slow everything to `$LINKSPEED` to prevent queuing at our ISP. Traffic class prioritization works only if you aren't being speed-limited by the router on the other end of the connection:

```
tc class add dev $DEV parent 1: classid 1:1 htb \
rate ${LINKSPEED}kbit burst 6k
```

Now, I'll add the first class of service, the premium one, as shown earlier:

```
tc class add dev $DEV parent 1:1 classid 1:10 htb \
rate ${LINKSPEED}kbit burst 6k prio 1
```

Here I'll the second class of service, the economy one, as shown earlier:

```
tc class add dev $DEV parent 1:1 classid 1:20 htb \
rate ${LINKSPEED}kbit burst 6k prio 2
```

Finally, I'll added the "default" class, as shown earlier. This is where undefined traffic will fall—i.e., anything not on the ports I specified in the previous section:

```
tc class add dev $DEV parent 1:1 classid 1:30 htb \
rate $[9*$LINKSPEED/10]kbit burst 6k prio 3
```

The following command makes IP packets that have the IP TOS header set to 0x19 match Class 1, our premium class. Remember that if you're prioritizing by port number, you might have no need to prioritize by the ToS header. So, you might be able to skip this line, especially if you have no control over your ToS headers, as discussed earlier:

```
tc filter add dev $DEV parent 1:0 protocol ip \
prio 10 u32 match ip tos 0x19 0xff flowid 1:10
```

The following line says that any packets with an IP TOS header equal to 0x18 will match class two, the economy class. This isn't foolproof either; not all packets even have a ToS header value inserted. Starting to get the idea? Discriminating by port numbers is more consistent and easier to manage than discriminating by ToS bits, but there it is, so you can see how it's done:

```
tc filter add dev $DEV parent 1:0 protocol ip \
prio 20 u32 match ip tos 0x18 0xff flowid 1:20
```

The simple loop shown in the following snippet makes sure that the ports defined in the $CLASS1PORTS variable match Class 1:

```
for a in $CLASS1PORTS
do
        tc filter add dev $DEV parent 1:0 protocol ip \
          prio 11 u32 match ip dport $a 0xffff flowid 1:10
        tc filter add dev $DEV parent 1:0 protocol ip \
          prio 11 u32 match ip sport $a 0xffff flowid 1:10
    done
```

Likewise, the simple loop in the following snippet makes sure that the ports defined in the $CLASS2PORTS variable match Class 2:

```
for a in $CLASS2PORTS
do
        tc filter add dev $DEV parent 1:0 protocol ip \
          prio 24 u32 match ip dport $a 0xffff flowid 1:20
        tc filter add dev $DEV parent 1:0 protocol ip \
          prio 24 u32 match ip sport $a 0xffff flowid 1:20
    done
```

The following code says that any traffic not matching the other rules is "bulk" and ends up in the bulk class, neither economy nor premium:

```
tc filter add dev $DEV parent 1: protocol ip \
prio 30 u32 match ip dst 0.0.0.0/0 flowid 1:30
```

For any type of commercial service, you will certainly want to work on this script a bit. Smart users could cheat you by hacking their IP ToS headers or by using a different port number to "stow away" in the premium class. So, regardless of how you implement classes of service, you'll also need to ensure that users of your service are authenticated with usernames and passwords, and you'll need to emphasize good logging so that you always know who's using which levels of service and so that nobody breaks the rules.

—*Kristian Kielhofner*

HACK #79 Build a $100 PSTN Gateway in 10 Minutes or Less

The Sipura SPA-3000 is a marvel of engineering. For less than a hundred bucks, you can interface your phone line with your VoIP network. What's cooler than that?

When your phone line is connected to your plain-old analog phone, it works and works well. But when it's connected to your VoIP network, it takes on a whole new personality. Suddenly you can do all kinds of cool stuff.

This hack will show you how to connect a PSTN phone line to an Asterisk-based VoIP network using the Sipura SPA-3000 ATA. This device is like other ATAs in that it has one FXS port. However, the SPA-3000 has a trick up its sleeve: a single FXO port as well. Not only does it have the hardware, but Sipura's firmware is actually quite flexible, allowing you to do all kinds of things to impress your friends and make life easier (hopefully). For this hack you'll need an Asterisk machine nearby [Hack #4].

I am going to demonstrate this hack using Asterisk and the SPA-3000. But because the SPA-3000 speaks SIP, you can just as easily use it in conjunction with most other SIP-compatible devices out there. (In fact, the Clipcomm CG-200 gateway [Hack #43] would make a fine substitute.)

The Asterisk server has a *sip.conf* file that allows calls to be placed into the default context from remote SIP endpoints. I am going to assume that you want incoming calls to the FXO port on the Sipura to be forwarded to an extension on that existing Asterisk server; I'm using 1000 for this hack. I'm also assuming the Sipura and the Asterisk server are on the same LAN and the server's IP address is 192.168.1.1.

On your Asterisk server, open up */etc/asterisk/sip.conf* and create a new entry at the bottom of the file:

```
[spa3k]
type=friend
username=spa3k
secret=spa3k ;<----- Pick a new password and write it down!
dtmfmode=rfc2833
host=dynamic
context=default
nat=yes
allow=all
```

Save *sip.conf* and reload Asterisk with **asterisk -rx reload**. If you would like to place outbound calls using your new SPA-3000, continue reading. Otherwise, you can skip ahead to the section "Configuring the Sipura." Next we will need to edit */etc/asterisk/extensions.conf*. Underneath the [globals] section, add a new line:

```
TRUNK=SIP/spa3k
```

If you already have a TRUNK variable defined, it is up to you to figure out how you want to mix and match your existing trunk(s) with your SPA-3000. Now, scroll down to the bottom of the file and add a new section here:

```
[spa-trunk]
exten => _NXXXXXX,1,Dial(${TRUNK}/${EXTEN},20)
exten => _NXXXXXX,2,Congestion

exten => _NXXNXXXXXX,1,Dial(${TRUNK}/${EXTEN},20)
exten => _NXXNXXXXXX,2,Congestion

exten => _1NXXNXXXXXX,1,Dial(${TRUNK}/${EXTEN},20)
exten => _1NXXNXXXXXX,2,Congestion

exten => _011.,1,Dial(${TRUNK}/${EXTEN},20)
exten => _011.,2,Congestion

exten => _NXX,1,Dial(${TRUNK}/${EXTEN},20)
exten => _NXX,2,Congestion
```

This dial plan will enable NANPA-style dialing of local, 10-digit local, long-distance, international, and emergency/information services from your system to the SPA-3000. You will want to make sure to include this new section in your local phone configuration. So, if your SIP phones, as defined in *sip.conf*, are in the "local" context, you will want the local context in *extensions.conf* to contain this line:

```
include => spa-trunk
```

This will enable your SIP phones to use your new PSTN gateway. Save *extensions.conf* and reload Asterisk with **asterisk -rx reload**.

Configuring the Sipura

Once you have unpacked the Sipura, connect your POTS telephone line to the RJ11 jack labeled LINE, connect an analog telephone to the RJ11 jack labeled PHONE, connect Ethernet, and then power up. Once the Sipura has powered up, dial **** from the analog telephone. As soon as you hear the voice prompt, dial **110#**. The answering voice will read back the SPA-3000's IP address.

Moving to your PC, enter the SPA-3000's IP address in your web browser. You should see a gray screen with some status information. In the upper right-hand corner, click Admin, and then click Advanced. You should see a wealth of new options appear.

Move over to the PSTN Line tab. Table 6-1 shows you the values to fill in for this page.

Table 6-1. Values to place in the PSTN Line configuration

Field name	Value
Proxy	IP address of Asterisk server
Username	spa3k
Display Name	spa3k
Password	spa3k
Register	Yes
Make Call without Register	No
Ans Call without Register	No
Dial Plan 8	(S0<:1000)
PSTN Ringthrough	No
PSTN Default DP	8
PSTN Answer Delay	8

After you have entered these changes, click "Submit all changes." The Sipura will reset, and once it reboots, you should have a fully functioning SIP/PSTN gateway, connecting calls between your Asterisk server and the PSTN.

—*Kristian Kielhofner*

Make IP Phone Configuration a Trivial Matter

#80

Trivial File Transfer Protocol (TFTP) servers are simple, stripped-down file storage servers that play a role in VoIP networks that's anything but trivial.

When an IP phone is first powered up, it goes through a boot-up sequence that's similar to that of a PC. While most PCs boot up from a functional configuration that's stored locally, an IP phone's configuration can be controlled remotely by an administrator, who can store each configuration in a central repository. IP phones use the TFTP protocol to retrieve updated configurations from that repository, known as a *TFTP server*.

With a TFTP server, you can centrally store and manage an entire network's worth of IP phone configurations. Merely placing a particular phone's configuration file on the TFTP server will change the phone's functionality to match the new configuration file the next time the phone is booted. Many ATAs can be configured this way, too. In addition, firmware updates for IP phones and ATAs can be delivered by TFTP. So clearly, understanding a little about TFTP and learning how to set up such a server is useful for any VoIP network.

Set Up a TFTP Server

TFTP servers can be hosted on Windows, Mac, Linux, and BSD machines (and on Cisco routers, too), and there's a host of free software (and plenty of good shareware) to enable a basic TFTP server for your test lab. For a more robust TFTP server, use the age-old *tftpd* software that's included in most Unixes.

To set up a simple TFTP server on Windows, download TFTP Desktop from Weird Solutions. This is a limited version of Weird's commercial Windows TFTP server, and it allows one transfer at a time—enough to satisfy the needs of a VoIP network with a half-dozen IP phones or so. (You can find TFTP Desktop at *http://www.weirdsolutions.com/weirdSolutions/pages/02products/download/index.htm.*)

Mac users can turn to my favorite TFTP server, Fabrizio La Rosa's aptly named TFTP Server, available at *http://www.macupdate.com/info.php/id/11116.* Like TFTP Desktop for Windows, this application creates a basic TFTP server with a simple graphical user interface (GUI), but it does so using Mac OS X's built-in TFTP daemon, meaning there are no limits to how much traffic it can handle. You just tell TFTP Server what folder you want to share using TFTP, and that's it.

Linux and other Unix users can probably find a fully functional TFTP server already on their systems. To be sure, issue a **tftpd &** at the root command line. If you don't get a "command not found" response, you've got the TFTP daemon, and you just have to figure out how to make it work. Thankfully, there's nothing to it. Just create a directory to create your TFTP repository (commonly */tftpboot*), and launch the TFTP daemon like this:

```
# mkdir /tftpboot
# chown nobody /tftpboot
# tftpd -l -s /tftpboot &
```

This will launch the TFTP server (which on many Linux systems runs as the phantom user *nobody* for security reasons—hence the chown to give *nobody* permission to the TFTP directory). Any files placed in */tftproot* will be accessible from a TFTP client, such as the one built into IP phones.

Understand IP Phone Configuration

The syntax of the phone configuration files saved on the TFTP server varies from one make of IP phone to the next. That is, Cisco has a different configuration-file structure than Uniden, for example. But they all bear a few things in common.

First, most IP phone configuration files are regular text files that look like a Unix *.conf* file or an old-fashioned Windows *.ini* file. Second, most IP phones support two methods of configuration by TFTP: default and phone-specific.

Default configuration files apply to all phones of a specific make that connect to the TFTP server. Phone-specific configuration files apply only to an individual IP phone. These tend to be denoted by the IP phone's MAC hardware address in the filename of the configuration file. So, if you were to browse through the contents of a typical TFTP server on a VoIP network, you'd see at least one filename without a MAC address (the default configuration) and a handful of files that have MAC addresses in their filenames, denoting them as phone-specific configurations. Most IP phones will revert to the default configuration automatically if no phone-specific file exists on the TFTP server. Other phones will include settings from the default configuration file if they *aren't mentioned* in the phone-specific file.

Uniden's IP phones generally follow this convention. Take a look at this sample */tftpboot* directory:

```
uniden00e01102ffb7.txt   uniden00e01103000a.txt   unidenBase.txt
uniden00e01102ffc3.txt   uniden00e01103001c.txt   unidencom.txt
uniden00e01102ffca.txt   uniden00e01103001f.txt
```

This directory contains several phone-specific configuration files (the *unidenXXXXXXXXXXXXX.txt* ones) and a default configuration file, *unidencom.txt. unidencom.txt* covers the networking configuration of the phones (specifically, SIP proxies). So, if you wanted a particular config to affect all the phones (unless overridden in the phone-specific configs), you would place that config in *unidencom.txt.* Cisco, Grandstream, and others use a very similar model to Uniden. For some tips on Cisco and Uniden TFTP configuration, check out the hacks in Chapter 5. (The *unidenBase.txt* file is there just as a template from which to generate future phone-specific config files.)

HACK #81 Peek Inside of SIP Packets

When the going gets tough, the tough sniff packets.

Adding VoIP to any network can be a daunting challenge, but accomplishing the task can seem particularly impossible when problems begin to arise. To help troubleshoot any network problems, being proficient with the use of network-analysis tools can provide some restful nights. Ethereal is a network-analysis tool that allows for the "sniffing" or capturing of data packets on the network. Ethereal has some VoIP-specific features, too. By digging deep into VoIP signaling conversations with Ethereal and assessing SIP traffic problems using a conventional call-flow graph, you'll reveal the source of many problems you're likely to encounter. Because of the dominance of SIP, this exercise will concentrate on using only the features of the Ethereal sniffer that support analyzing this protocol rather than any of the protocols that came before it (MGCP, H.323, etc.). Feel free to learn those other protocols if you like, but know that almost every commercial VoIP provider uses SIP (when I say *almost*, I mean 95% or more).

You can use Ethereal to inspect network traffic from the Ethernet layer all the way up to the application layer. Packets are captured in a buffer and are displayed on the screen. Filters can be applied to restrict the capture to packets matching a certain source, destination, size, protocol, or service.

To obtain Ethereal, download it from *http://ethereal.com/* and install it on your Windows, Mac, or Linux box. The screenshots and examples here assume the Windows version.

On a switched network, a nonadministrative user can only capture packets being sent to or from his own machine. So, to keep things simple, this hack monitors traffic for the SIP softphone known as X-Lite [Hack #4]. Both X-Lite and Ethereal will need to be running on the same machine. If you're using a nonswitched network, like a hub, Ethereal can observe packets not bound for or originated from your PC, which means you'll be able to monitor all

VoIP traffic without restrictions. But for now, I'll assume you can monitor traffic only on your own PC.

> Another way to view packets bound for other PCs on a switched Ethernet network is to use a technique known as *port spanning*. With port spanning, you can program an Ethernet switch to let you snoop traffic on ports other than the one where your PC is connected. Check out *Cisco IOS in a Nutshell* (O'Reilly) for an introduction to port spanning.

To demonstrate SIP packet observation with Ethereal, we'll set up a filter that allows us to capture SIP registration signals in two scenarios: one for a successful SIP registration and another for a failed SIP registration. As in the other projects, the SIP server's IP address is 10.1.1.10. In this instance, Asterisk is used as the SIP server.

> X-Lite offers excellent diagnostic logging, too. Some of the packets you observe with Ethereal in this project will correlate with entries in the X-Lite diagnostic log, which you can view by selecting Diagnostics from the right-click menu in X-Lite's UI.

Configure the SIP Softphone

If you're setting up X-Lite for the first time, you'll need to click the Configuration button, right of the center, next to the Clear button (see Figure 6-2). Once you click this button, you'll see the Configuration menu.

In the Configuration menu, double-click Menu and then select System Settings → SIP Proxy → [default]. This will take you to the SIP client configuration, as shown in Figure 6-3. Here, you can configure the softphone to register using a number and/or password to match what you've established in the dial-plan configuration on the SIP server.

Once configured, the SIP softphone will automatically register with the SIP registrar as soon as you close the Configuration menu. If registration was successful, you'll see Logged In in the UI display, as shown in Figure 6-2. If it wasn't, make sure the SIP proxy profile called [default] is enabled and is configured to match a SIP account on the server.

Configure Ethereal

Once Ethereal is installed, launch it. Next, begin a capture by selecting Capture → Start. This will show you the Capture Options dialog, shown in

Figure 6-2. The X-Lite softphone's user interface

Figure 6-4. To limit the kind of traffic that Ethereal will capture, you'll need to use a filter string. Ethereal has a rather sophisticated syntax for this string, which instructs Ethereal what to capture and what to ignore. This syntax is explored more in *Managing Security with Snort and IDS Tools* (O'Reilly).

In this case, our SIP server is 10.1.1.10, and the standard port for SIP traffic is UDP 5060. We want to capture traffic in both directions—that is, to the SIP server and to the softphone running on the same host as Ethereal. Here is the string that achieves this:

```
host 10.1.1.10 and udp port 5060
```

Check the "Update packets in real time" and "Automatic scrolling in live capture" options to see the packet capture log occur immediately instead of waiting until the capture session is complete. Then, click OK, and the main capture window will appear. You're now ready to observe your SIP registration attempts.

Observe SIP Registration

Now, restart X-Lite. It will attempt to register automatically with the SIP server upon startup. By the time X-Lite says you're "Logged In," you can stop the packet capture by clicking Ethereal's Capture → Stop menu item.

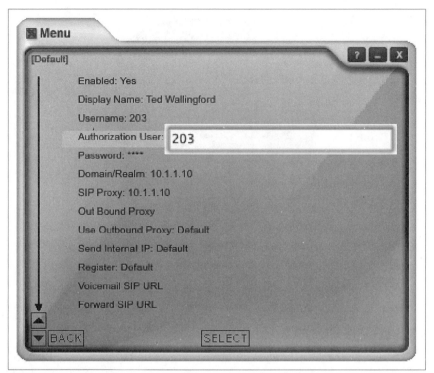

Figure 6-3. X-Lite's SIP client configuration

The main capture window should be filled up with a number of packets, as shown in Figure 6-5.

In this instance, Ethereal shows the first packet, packet number 1, as a SIP REGISTER method. Newer versions of Ethereal, such as the 0.10.7 version used here, can parse SIP packets and tell you which methods and responses they contain. Packet 4 is a second registration request (the first one failed because X-Lite tries anonymous registration first).

Packet 5 is the 100 Trying SIP response sent from the SIP server back to the softphone. Packet 6 is the 200 OK SIP response sent from the SIP server back to the softphone, indicating that the registration was successful. Packet 7 is a SIP NOTIFY method asking for username 204 at the SIP registrar. Packet 8 is the 200 OK response. At this point, registration is complete. The additional packets (9 and 10) are keep-alive packets that X-Lite sends to its SIP registrar. Not all SIP phones do this.

The bottom pane of the main capture window shows the actual hex-encoded content of the packet and the ASCII-encoded content of the packet

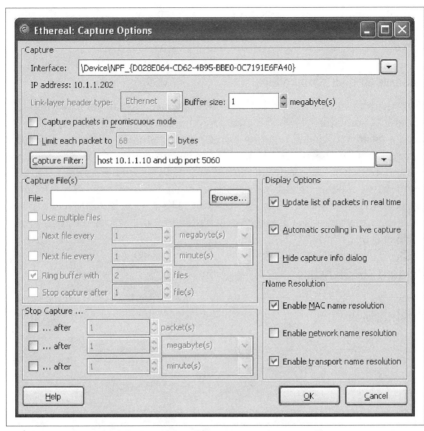

Figure 6-4. Ethereal's Capture Options dialog

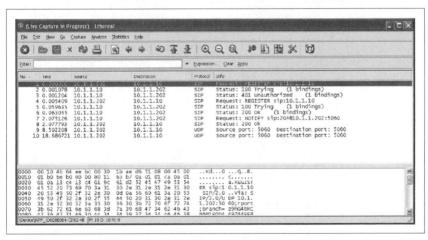

Figure 6-5. An Ethereal capture of a SIP registration

that corresponds to it. The hex is on the left, and the ASCII is on the right. This is where you can usually pick out problems: an incorrect password or a botched username would be easy to spot this way.

Observe Registration Failure

Outside the test lab, you probably won't have occasion to capture SIP packets unless something is working incorrectly. But then, why else would you need a packet sniffer? Simulating a registration failure is very easy. Just alter the registration username of the SIP proxy profile in X-Lite to one that doesn't match a SIP peer on the SIP server. (Be sure that your softPBX requires SIP clients to use a username and password; you can configure Asterisk to allow anonymous registration!) Then, start the Ethereal capture with the filter string used earlier, restart X-Lite, and watch the registration crash and burn.

Capture SIP Statistics

Once you have saved a capture file, you should be left with a trace screen that looks something like Figure 6-6, showing all the data packets captured.

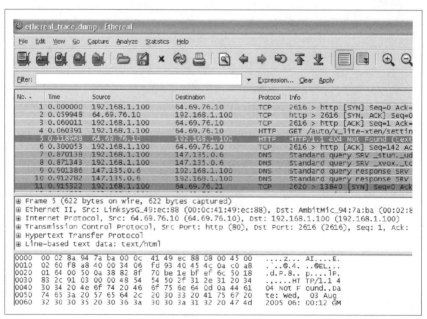

Figure 6-6. An Ethereal packet capture log

To help sort through all this captured data, you will use Ethereal's handy SIP statistics tool by navigating to Statistics → SIP. This tool produces a simple report on the statistics of the SIP messages sent and received on the interface, as shown in Figure 6-7.

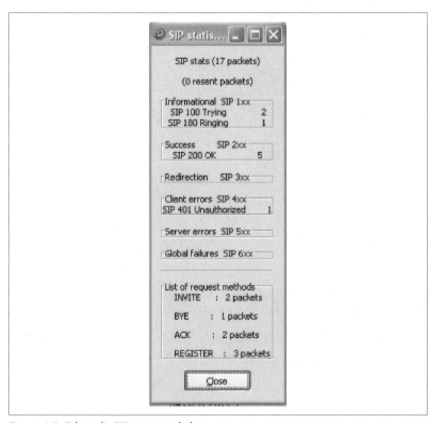

Figure 6-7. Ethereal's SIP statistics dialog

Ethereal has many built-in tools for filtering and colorizing data traces, but it can be a bit overwhelming to digest all of the data at once. To help network administrators graphically view the call flow of a SIP signaling conversation, navigate to Statistics → VoIP Calls. You are presented with a list of captured voice calls, from which graphs like the one in Figure 6-8 can be produced.

The illustration technique used in Figure 6-8 is a standard way of representing data conversations. In Chapter 7 of O'Reilly's *Switching to VoIP*, the details of SIP signaling are covered with a half-dozen examples that are graphed in this fashion. A standard convention of a SIP call-flow graph is the direction arrow. In this example, there are many of them: one for each SIP message sent. When you click on one of the arrows, it automatically drills

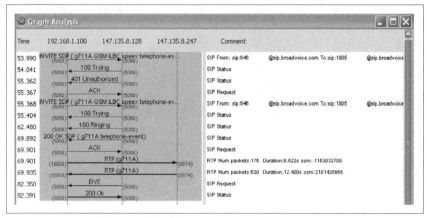

Figure 6-8. A SIP call, graphed by Ethereal

down to the specific packet regarding that part of the call flow in the trace window. This is fantastic for literally stepping through the signaling steps of call setup and tear-down.

—Joel Sisko

HACK #82   Dig into SDP

No SIP-based telephone call happens without Session Description Protocol, and there's an SDP message inside every SIP Invite.

Knowing what SDP messages look like will help you identify the cause of failed calls that are rooted in incompatible phones. That's right; if the phones (or softphones) calling each other with SIP don't have (or aren't configured to use) the same media codecs, they won't be able to talk to each other! Of course, the phone isn't going to say, "Hey dude, I don't have the right codec!" It's more than likely just going to give you a busy signal. So you'll need to dig into SDP to find out if a mismatch of media capabilities is to blame.

SDP is an essential part of SIP call signaling. Its elements are text tokens sent in SIP packets with the SDP content-type header. These tokens advertise the capabilities and requirements of each endpoint according to the parameters of the application, be it a telephone call, instant message, or something else.

During call setup, specifically during the SIP INVITE method, the SDP payload is sent from one endpoint to the other. A SIP 200 OK response indicates agreement with the SDP parameters, and a 4xx response indicates disagreement or incapability. For a much deeper discussion on SIP, have a peek inside my book, *Switching to VoIP* (O'Reilly).

Inspect Successful Capabilities Negotiation

Using Ethereal configured with the same filter string from "Peek Inside of SIP Packets" [Hack #81], you can capture a successful capabilities negotiation. In its default configuration, Asterisk supports G.711 so that just about any IP phone, including X-Lite, can place calls to it. In this case, X-Lite will be used to call Asterisk extension 201, and the SDP exchange for this call will be captured.

If you don't have such an extension on your dial plan, you can call Asterisk's default auto-attendant demo at extension 500 instead. (If you've removed this extension in your hacking of Asterisk, just run a make samples from your Asterisk source directory [Hack #41] to get the default config back again.)

When you place the call on X-Lite, use Ethereal to capture the SIP packets and zero in on the SDP content carried in the INVITE methods and 200 OK responses. In Figure 6-9, you can see that the call setup was successful.

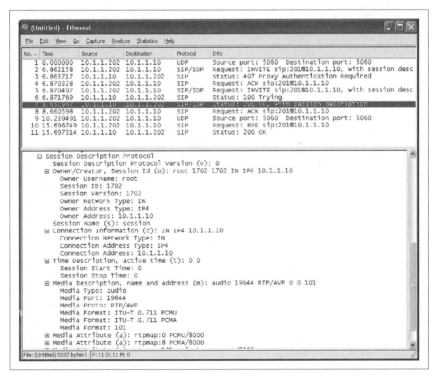

Figure 6-9. Ethereal can parse SDP content so that it's easier for you to troubleshoot call setup problems

Packet 5 is the authenticated INVITE method. The user in this example is calling SIP user 201. Included in packet 5 is an SDP payload. Ethereal indicates this in its Protocol column in the top packets pane of the main capture window, shown in Figure 6-9. Packet 6 is the 100 Trying response. Packet 7 is the OK response, which also includes an SDP payload. If there's a codec match in the media attributes list of the SIP INVITE and the 200 OK response (shown in the bottom pane), all that's needed is a SIP ACK method sent by the caller to confirm agreement on the first matching attribute. That's what packet 8 is.

Inspect Failed Capabilities Negotiation

If there was no capabilities match, call setup would fail. This scenario can be produced easily by temporarily crippling the capabilities of the Asterisk server. To make it impossible for the X-Lite softphone to negotiate an audio stream with the Asterisk server, you can disallow all codecs supported by X-Lite and permit only Global System for Mobile (GSM) codecs, which X-Lite doesn't support. Take a look at this snippet of /etc/asterisk/sip.conf, which does just that:

```
[general]

port = 5060         ; Port to bind to (SIP is 5060)
bindaddr = 0.0.0.0  ; Address to bind to (all addresses on machine)
disallow=all
#allow=ulaw
#allow=alaw
#allow=gsm
```

The G.711 and GSM codecs have been commented out. This simulates a codec capabilities mismatch, so the SIP client won't be able to pass the SDP negotiation and call setups will fail. (Don't forget to issue a "reload" command at the Asterisk command line.) Now, we can run the capture again while X-Lite tries to call extension 500. (In Ethereal, select Capture → Start.) Only this time, the call will fail because there's no suitable codec common to both the caller (X-Lite) and the receiver (Asterisk).

Packet 7 is shown in Figure 6-10. It's a SIP INVITE carrying SDP content that includes a list of a tokens. These represent media attributes, or capabilities. Ethereal presents the SDP content in a parsed, hierarchical fashion.

The raw ASCII SDP payload of this SIP packet, which can be seen in X-Lite's diagnostic log, actually looks like this:

```
...
Content-Type: application/sdp
User-Agent: X-Lite release 1103m
Content-Length: 290
```

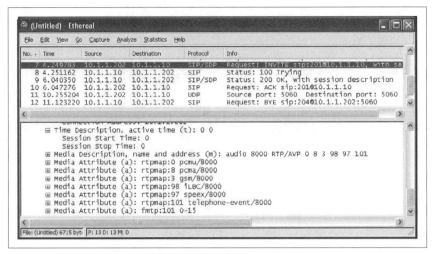

Figure 6-10. The bottom pane of the capture window shows the media attribute list: the SDP text payload that advertises the capabilities of the calling endpoint

```
v=0
o=203 146336832 146337009 IN IP4 10.1.1.201
s=X-Lite
c=IN IP4 10.1.1.201
t=0 0
m=audio 8000 RTP/AVP 0 8 3 98 97 101
a=rtpmap:0 pcmu/8000
a=rtpmap:8 pcma/8000
a=rtpmap:3 gsm/8000
a=rtpmap:98 iLBC/8000
a=rtpmap:97 speex/8000
a=rtpmap:101 telephone-event/8000
a=fmtp:101 0-15
```

The capabilities are listed with a reference number following the rtpmap key-word. 0 pcmu/8000 indicates that 0 is the reference number that RTCP will later use to refer to this G.711 uLaw at 8000Hz capability. The other capabilities are advertised with other numbers. (These numbers are reserved like commonly used port numbers in TCP/IP, and they can be overridden.)

In Figure 6-11, you can see that the 200 OK response sent by the receiver to the sender has an SDP payload that presents no audio codecs at all in its media attributes. This is because they have been purposefully disabled, of course. Packet 10 is the customary SIP ACK method acknowledging receipt of the 200 OK and giving the go-ahead for RTP to begin. But without any matching SDP media attributes to establish the RTP media channel, the receiver selects attribute reference number 101 using SDP's m token. 101 means that no valid capabilities match. RTCP will report to the calling endpoint a few seconds later that no media channel exists, and the receiver "hangs up" with a SIP BYE method in packet 12.

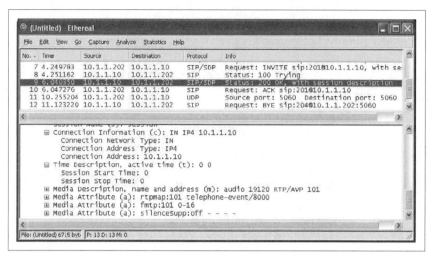

Figure 6-11. The bottom pane of the capture window shows the media attribute list: SDP's listing that advertises the capabilities of the receiving endpoint, in this case Asterisk

 Don't forget to re-enable the codecs after doing this experiment, or you'll have a *real* problem to troubleshoot!

Call-signaling issues can be frustrating, especially when using a mixed bag of SIP products from different vendors and vintages. Just like for revealing SDP failures and authentication problems, packet capture is the best tool for exposing any and all signaling problems.

 ## HACK #83 Sniff Out Jittery Calls with Ethereal
I have seen the enemy, and its name is Jitter.

One of the biggest concerns when using VoIP is packet jitter. Jitter is the difference in time that packets take to arrive at the final destination. The greater the difference, the worse your calls will sound, and the more you'll want to hang up that phone and return to a traditional telephone system. But it doesn't have to be that way. Instead of throwing your hands up and giving up on the converged network dream, you can just get serious about the jitter problem, and the first step is identifying instances of jitter on your network.

You've seen how to do this in a "big-picture fashion" using *RRDtool* and *sip_ping.pl* [Hack #75], but that technique has a few shortcomings. Though it gives you a long-term assessment of jitter conditions on the network, it doesn't do so using RTP, the protocol that carries real voice payloads, and it takes samples only every five minutes, meaning that you can't assess the jitter conditions for any given call. This is where Ethereal can really help you out.

Identify Jitter

If you've looked at "Graph Latency and Jitter" [Hack #75], you've already seen the face of the enemy. With some help from Ethereal, you can zero in on jitter and prepare to squash it like a bug.

When you're examining jitter, you're mainly concerned with RTP packets. To use Ethereal here, you must first locate an RTP packet in the trace file screen. (You'll need to have grabbed a packet sniff like the one you grabbed in "Peek Inside of SIP Packets" [Hack #81].) Once you find an RTP packet in the trace file screen, navigate to Statistics → RTP → Stream Analysis. Figure 6-12 shows the report analysis.

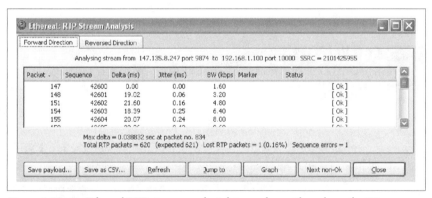

Figure 6-12. An Ethereal RTP stream analysis lets you know if you have the jitters

By examining the stream analysis, you can see that, at least in this particular sequence, the jitter is nearly nonexistent, staying well below a rate of 1 ms. In a problem scenario, jitter would need to be 10 to 20 ms at a minimum to be audibly perceived (of course, the codec has a lot to do with how much jitter the human ear can tolerate; G.711 is highly resilient to jitter, and G.729 is less so).

The Jitter Solution

Once you've got the jitters, the only way to get rid of them is to implement QoS at the points on your network from which jitter is originating. Typically, these are routers, VoIP servers, or extremely busy switches. More than a dozen different QoS specifications are available (among them DiffServ, RSVP, VLAN, and IP Precedence, to name a few), and there are probably twice that many ways to implement them. So how do you choose?

Good question. There's no simple answer, because each specification is aimed at a different solution. RSVP is a bandwidth reservation technique for WANs, and Virtual LAN (VLAN) is a traffic-segmentation technique for LANs. Both have implications for QoS that are deserving of their own book (I like the hardcover classic, *Quality of Service* (Cisco)).

Jitter might in fact be a losing battle, depending on how you use VoIP and where your VoIP calls travel. If they go across a network jurisdiction that's out of your control, like the Internet, it might be impossible to provision QoS, and you might never get acceptable voice quality. The moral of this story is very basic: you can perfect VoIP quality on private, controlled networks, but on the Internet, it's a crapshoot.

—Joel Sisko

HACK #84   Log VoIP Traffic

A Linux PC's built-in IP router and firewall, NetFilter, can be a useful tool for logging VoIP traffic.

In a scenario where several satellite offices on a WAN (or the Internet) are linked together as an IP telephony network, origin- and destination-based logging is crucial, because it will tell you which office is using the most VoIP capacity, which is using the least, and when it's all being used.

When a Linux NetFilter firewall is used to protect a group of enterprise VoIP servers or just as a gateway router for a segment where VoIP is used, a lot of VoIP-related events can be monitored and logged. Logging from the firewall is useful for the security-minded, but it's important for other reasons, too. It lets you get a feel for which remote networks and hosts are communicating with your VoIP services and how often they are doing this. This will improve your understanding of bandwidth consumption and traffic patterns on your network, besides giving you a keener awareness of security.

Logging with NetFilter

NetFilter's default configuration provides for no logging. If you want a particular type of packet logged—say, from a specific network or on a specific port—you must tell NetFilter to log it. When a packet is logged, its pertinent information is sent to *syslog* to be stored. *syslog* is the system-wide logging daemon that is a staple in most Unix-variant operating systems.

> Logging packets using NetFilter doesn't save the contents of the packets—just information from the packets' headers! If you want to *capture* packets, you'll need other software, like Network Associates' Sniffer or the open source tool Snort.

To enable logging, you must set up a rule that specifies which packets you want to log. The following rule says to log all packets sent *to* the machine running the NetFilter firewall (keep in mind that this will eat up tons of storage space *fast!*):

```
# iptables -A INPUT -j LOG --log-prefix "Log it all baby."
```

The log prefix options allows you to specify what will appear at the beginning of the log entry for each packet. That way, when you comb through lengthy databases of these entries, you can find specifically what you're looking for. The following rule is very broad; it captures any and all SIP traffic going *through* the firewall (FORWARD chain) and logs it:

```
# iptables -A FORWARD -p udp --dest-port 5060,5061 -j LOG \
--log-prefix "SIP"
```

Let's say that you are operating a SIP proxy that facilitates VoIP calling via SIP directly to two other proxies. Let's also say that all three SIP proxies are in the same organization and that a site-to-site VPN is used to connect them all. The three proxies support three VoIP LANs at separate offices. The LANs they support have the network addresses 10.1.0.0/16, 10.2.0.0/16, and 10.3.0.0/16, as in Figure 6-13. The configuration examples given in this section are assumed to be running on the firewall in the 10.1.0.0 network.

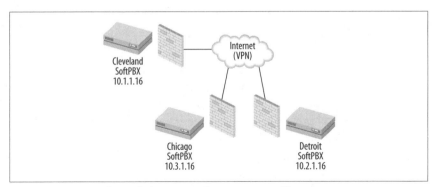

Figure 6-13. Three physically separate softPBXs connected by an Internet VPN

Assuming a VoIP WAN like the one in Figure 6-13, it's possible to do some interesting logging and prefixing. Say you want to log SIP traffic by remote network. You can use the following commands to tag inbound and outbound traffic:

```
# iptables -A FORWARD -P udp -s 10.2.0.0/16 --dest-port 5060 -j \
LOG --log-prefix "FromDetroit"
# iptables -A FORWARD -P udp -d 10.2.0.0/16 --dest-port 5060 -j \
LOG --log-prefix "ToDetroit"
# iptables -A FORWARD -p udp -s 10.3.0.0/16 --dest-port 5060 -j \
LOG --log-prefix "FromChicago"
# iptables -A FORWARD -p udp -d 10.3.0.0/16 --dest-port 5060 -j \
LOG --log-prefix "ToChicago"
```

This example tags all the Detroit traffic separately from the Chicago traffic, making it easier to discern later on when you're viewing the packet log.

A simple modification to the previous example would allow you to log RTP traffic (port 8000 or whatever your endpoints use). On a strategically placed Linux firewall, this could provide valuable information about bandwidth consumption at the Detroit and Chicago sites in the example.

Another technique is to differentiate VoIP traffic that is to/from the private network from that which is to/from the Internet or another foreign VoIP network:

```
# iptables -A FORWARD -p udp -s 10.0.0.0/8 --dest-port 5060,8000 \
-j LOG --log-prefix "Private VoIP"
# iptables -A FORWARD -p udp -d 10.0.0.0/8 --dest-port 5060,8000 \
-j LOG --log-prefix "Private VoIP"
# iptables -A FORWARD -p udp -s 0.0.0.0/0.0.0.0 --dest-port 5060,8000 \
-j LOG --log-prefix "Internet VoIP"
# iptables -A FORWARD -p udp -d 0.0.0.0/0.0.0.0 --dest-port 5060,8000 \
-j LOG --log-prefix "Internet VoIP"
```

In this example, private traffic—that is, traffic to or from a 10.x.x.x host—is tagged separately from Internet traffic, which is indicated by the catch-all network address 0.0.0.0.

In a VoIP network that's connected to the Internet but doesn't use the Internet as a call path, it would be a good idea to log all VoIP traffic originating from the Internet. Such traffic could be an indicator of system abuse, just as email systems are abused by spammers. Logging this type of traffic would tip you off to somebody trying to originate long-distance phone calls from your system—something I've encountered in the field myself.

To dig into your VoIP traffic, consult Table 6-2, which is a list of commonly used VoIP port numbers, by protocol.

Table 6-2. Commonly used VoIP ports

Service	Standard port numbers
SIP	5060, 5061 (usually UDP)
H.323	2099, 2517, and 2979 UDP and TCP
MEGACO/H.248	2944 and 2945 TCP and UDP
TFTP	69 TCP and sometimes UDP
Asterisk Manager API	5038 TCP
RTP	Varies often among 5000, 5001, 10000, and 10001; always UDP
IAX	5036 and 4569 UDP

Read and Analyze VoIP Traffic Logs

Once you've used *iptables* to tell NetFilter to catch some VoIP traffic and log it, the log data is stored in the kernel facility of a *syslog* database file, where it can be retrieved using the dmesg command on Red Hat Linux. Other operating systems might provide different tools for viewing system logs.

When packets are logged, several bits of information about each packet are stored:

- The protocol of the packet (UDP, TCP, ICMP, etc.)
- The date and time the packet traversed the NetFilter chain where it was logged
- The size of the packet
- The source and destination addresses and ports (sockets)
- The originating MAC hardware address (when the packet comes from an Ethernet interface)
- Of course, the prefix that you specify in the --log-prefix option, if any

dmesg, an application for reading kernel logging messages like the ones you captured earlier, provides flat text output that you can redirect to a file or pipe into another application for further filtering—like *grep*. Suppose you want to isolate the traffic prefixed with "Chicago" into a file by itself. You can use this command:

```
# dmesg | grep Chicago >chicagoVoIP.txt
```

Or, better yet, email that log to somebody, perhaps so that they can import it into a spreadsheet for further analysis. In the following example, hitting Ctrl-C will stop the dmesg application, and an email will be sent containing the "Chicago" entries:

```
# dmesg | grep Chicago | mail chicagoVoIPadmin@example.com
```

By combining good logging with good, high-level network traffic assessment [Hack #75], you'll have an excellent grip on what's happening beneath the application layer on your VoIP network. You'll know *how* it all works, because you'll have many techniques to observe VoIP packet flow, including the packet logging I've just described.

See Also

- "Graph Latency and Jitter" [Hack #75]
- "Peek Inside of SIP Packets" [Hack #81]
- "Sniff Out Jittery Calls with Ethereal" [Hack #83]

HACK #85 Secretly Record VoIP Calls

G.711 uLaw is the most common codec used in enterprise VoIP, but it's far from secure.

The G.711 codec is the de facto standard for voice encoding on VoIP networks, because the earliest VoIP gear and software didn't have enough processor power for real-time transcoding from one codec to another. This means that if a call were to originate on the PSTN and terminate on a VoIP device, the entire call would have to be in the same codec. The codec that's always been used on the (North American) PSTN is G.711 uLaw. Unfortunately, even as Cisco CallManager—arguably the world's first enterprise VoIP platform—became popular, it was painfully clear that running G.711 uLaw across the Internet was a very insecure thing to do.

That's because the RTP packaging convention used by most VoIP systems doesn't encrypt the media stream of a call, making it the aural equivalent of clear text, ripe for outside snooping. Using *tcpdump* and a copy of *vomit* (Voice over Misconfigured Internet Telephones), you can actually capture phone calls midstream and convert them into WAV files. How's that for security? (I argue that it's actually harder to secretly record calls with VoIP than it is on the PSTN, but let me digress here....)

To clandestinely record a G.711 uLaw phone call, you'll need to be able to run *tcpdump*, the common packet capture utility, or its Windows counterpart, *windump*. This means you'll need to be a privileged user on the machine you're going to record from (for Windows, this means Administrator; for Unix, it means root).

You'll also need the ability to view network traffic to and from the host(s) participating in the call. This means running the capture on one of the hosts directly, programming your switch to let you monitor the port where one of the hosts is connected, or (gasp!) connecting both hosts to a hub, where you can capture packets to your heart's content. To put this in plain English, unless you're using a hub or a specially configured switch, you'll be able to record calls only from a device that's actually on the call path—i.e., the caller's host, the receiver's host, or a VoIP server in the middle of the conversation.

The Hack

It's possible to do this hack on Windows (you'll need the same *WinPCap* library you used when you installed Ethereal on your Windows PC [Hack #81]; you did install Ethereal already, right?). However, I'll assume you're using Unix, since *tcpdump* is a standard Unix utility and because it's easier to install *vomit* on Linux or BSD than it is on Windows.

Compile and install libdnet and libevent. To download, compile, and install the *libdnet* and *libevent* libraries, required by *vomit*, log in as root and use these commands:

```
# cd /usr/src
# wget http://ufpr.dl.sourceforge.net/sourceforge/libdnet/libdnet-1.10.tar.gz
# tar xvfz libdnet-1.10.tar.gz
# cd libdnet-1.10
# ./configure
# make
# make install
# cd ..
# wget http://www.monkey.org/~provos/libevent-1.1a.tar.gz
# tar vzxf libevent-1.1a.tar.gz
# cd libevent-1.1a
# ./configure
# make
# make install
```

Obviously, this is just a sequence of commands to fetch the libraries, open the archives, and compile the source code within.

Compile and install vomit. Next, grab the *vomit* tarball and compile it on the same machine, again as root:

```
# cd /usr/src
# wget http://vomit.xtdnet.nl/vomit-0.2c.tar.gz
# tar zvfx vomit-0.2c.tar.gz
# cd vomit-0.2c
# ./configure
# make
# make install
```

tcpdump some packets. When I did this hack, I did it on my Asterisk server running on Linux. This simplified the capture process, since all I had to do

was set up an extension on the Asterisk server that answered the call imme-
diately and produced some audio. (For a refresher on doing this, flip back to
"Attach a SIP Phone to Asterisk" [Hack #42].) Once the extension was in place,
I started *tcpdump* like this:

```
# tcpdump -w test.file
```

When you use this command, it will create a dump file in the current directory that contains every IP packet sent or received by the default interface.
This file is going to get big pretty quick, so run this command only for as
long as is necessary to capture the call you're placing to the server. Then, at
the conclusion of the call, hit Ctrl-C to stop *tcpdump*.

"Wave" goodbye to privacy. Now, here's the truly fun part. The point of
vomit is to pick the G.711 RTP packets out of the dump file created by
tcpdump (*test.file*, as shown earlier) and string them together into a WAV
file. Try it:

```
# vomit -r test.file > test.wav
```

Run that WAV file through SoX if you need it in another format [Hack #24],
and off you go. Just don't record any calls without full knowledge of the participants, or you could find yourself in legal trouble.

HACK
#86   Log and Record VoIP Streams

Biblically speaking, there's not a whole lot that Cain and Abel had to do with
Voice over IP. But the program that bears their names is a really cool VoIP
tool.

If you're not a Unix fan or you just don't have the time to compile *vomit* and
its dependencies to record a call, I've got the solution for you. There's a program for Windows called Cain & Abel. It uses the *WinPCap* library (just like
Ethereal) for packet sniffing, network device identification, password recovery, reconnaissance, and literally dozens of other intriguing tasks. Cain &
Abel is literally a Swiss army knife of handy networking goodies.

Not least among these goodies is a VoIP call sniffer/recorder that's slicker
than a wet rock. It provides a sortable, date- and timestamped list view that
logs any VoIP calls it picks up during a sniff. It assumes that any RTP traffic
is VoIP and attempts to decode it and record it into a WAV file. According
to the Cain & Abel web site (*http://www.oxid.it/cain.html*), the program can
decode calls in uLaw, aLaw, ADPCM, LPC, GSM, iLBC, and a host of other
codecs. Of course, it can't interpret any streams that are encrypted, so it's
still nearly impossible to record a Skype call from another host.

Cain & Abel has a ton of password-cracking and network-ing-snooping stuff built in, so be sure to abide by the local policy of the network you're working on, or you could end up in a heap of trouble.

The Easy Way to Intercept Calls

To record a call from the local computer where Cain & Abel is running—that's the easiest way—install the program on a machine with X-Lite or a comparable softphone that can place calls in one of Cain & Abel's supported codecs. Of course, this technique will only allow calls placed to and from *this* machine. It will not sniff out calls between other computers or IP phones.

Fire up Cain & Abel. Then, select the Configuration menu option in Cain & Abel to launch the Cain & Abel configuration dialog. It's shown in Figure 6-14. Click the Filters and Ports tab and check the SIP/RTP entry to ensure that you'll be capturing VoIP traffic. Then click OK.

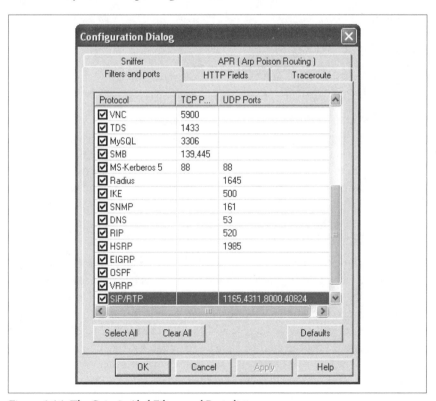

Figure 6-14. The Cain & Abel Filters and Ports list

When the Configuration dialog disappears, click the Start/Stop Sniffer icon on the toolbar. Now, place a phone call on the locally running softphone. This could be X-Lite, Firefly, NetMeeting, or whatever, as long as it uses SIP or H.323 for signaling and RTP for voice transmission (just about all VoIP applications do). Click the Sniffer tab, then the VoIP tab (on the bottom of the GUI) to reveal the call list.

Notice that as you place and receive VoIP calls on the machine where Cain & Abel is sniffing, your call log will begin to fill with entries on the VoIP tab, as shown in Figure 6-15. The call log will tell you the source and destination IP addresses of the media stream used in the VoIP call, the codec that is employed (if Cain & Abel recognizes it), and the port numbers involved in the RTP media path.

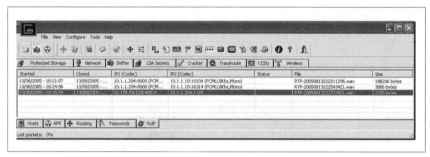

Figure 6-15. The Cain & Abel VoIP call log

After you stop the sniffer by clicking the Start/Stop Sniffer button on the toolbar again (it toggles sniffing on and off), you'll see the filenames where Cain & Abel has saved the recorded calls. The WAV files produced by Cain & Abel end up in \Program Files\Cain\VoIP, and you can play them by opening them in your favorite sound player, or by right-clicking them here in the Cain & Abel GUI.

The Tricky Way to Intercept Calls

If you want to record a call between two devices that can't run Cain & Abel, like a call between two IP phones or a call from a Mac softphone to a Linux softphone, the method described in the previous section won't work. Instead, you need to enable your Ethernet switch to "share" packets destined for the devices involved in the VoIP call with your PC running Cain & Abel. With your PC connected to a particular port, a typical managed Ethernet switch allows you to "listen in" on traffic on the other ports—like the ports where a VoIP call participant is connected.

On a nonswitched network, like a hub, you can use the "easy way" to monitor any device that's connected to the hub.

Cisco switches use a technique called Port SPAN to mirror the packets sent or received on one port to another port. In this manner, the switch administrator can inconspicuously capture all traffic on any port he chooses. To record a VoIP call, you'll need to set up port spanning between your PC's port and the target VoIP device's port. For the moment, I'm going to assume you're eloquent enough with Cisco configuration that you can at least get into your switch's command prompt and Enable mode. If you've no idea what this means, you might want to invest in James Boney's insightful *Cisco IOS in a Nutshell* (O'Reilly).

Let's say the VoIP device we want to record packets from is connected to port 5 on the switch. Use this command to mirror packets into what Cisco calls a "SPAN Session," a place we can retrieve them from on another port:

```
Switch(config)# monitor session 1 source interface fastethernet 5/1
```

Now, traffic to and from port 5 is mirrored to SPAN Session 1. Next, we need to reflect that traffic to the port where the sniffing PC is connected—say, port 4:

```
Switch(config)# monitor session 1 destination interface fastethernet 4/1
```

So now, traffic from port 5 will also occur on port 4, where the Cain & Abel PC can sniff it. (Don't forget a "write mem" if you want to keep the switch configured this way permanently.) Now you can use Cain & Abel (and *vomit*, for that matter) to record calls that traverse your switched network, even if you can't install a recorder on one of the participating VoIP devices.

Intercept and Record a VoIP Call
#87 This is the ultimate sneaky way to intercept and record VoIP calls.

I'm going to demonstrate why people say VoIP isn't secure. If you've got a laptop and a patch cable, you can record calls from a Cisco CallManager IP telephony network or even from a Vonage subscriber. More specifically, you'll be doing so without the need for port spanning on the switch [Hack #86] and without installing a recorder or sniffer on the device you're trying to record [Hack #85].

Instead, you'll be resorting to a tactic that's, well, unnatural. In fact, if you do this outside the test lab, it could be considered unethical, too, so be careful. I am about to teach you how to secretly listen in on other users' VoIP calls without having any direct contact with their VoIP phones or PCs and without being the administrator of the local network. Be advised, though, I'm not recommending that you ever do this in the field. I'm just passing on some knowledge I picked up while working as a networking consultant.

Get to Know Cain & Abel

While you're here, make sure you've read "Log and Record VoIP Streams" [Hack #86], which introduces the software tool I'll be using to make all this happen: Cain & Abel. If you don't know Cain & Abel, go back a hack, read it through, and you'll be able to proceed with confidence.

Now, on to the hack.

An Ethernet switch has anywhere from 4 to 48 ports where Ethernet devices like PCs and IP phones can connect. Each device connected to the switch has a 32-bit hexadecimal address, called a MAC address. Ordinarily, the switch knows which port to send a particular packet to because address resolution protocol (ARP) has been used by the sender (or by the closest router) to address the packet to the correct MAC address. All the switch has to do to get the packet to the right destination is transmit it on the port where the device with a matching MAC address is connected.

The secret to recording another person's VoIP calls is in making the switch think your MAC address is a valid destination for the VoIP traffic that's bound for that person's VoIP device. Specifically, when the sending device uses ARP to resolve the IP address of the intended recipient, your PC must respond by saying, "I am the holder of that IP address; send the data to me!" This hack, which rather goes against the prescribed way an Ethernet LAN is supposed to work, is called ARP poisoning. The result is that your PC intercepts the packets destined for the intended MAC address, so you can do with them whatever you like.

Once intercepted, the packets must be forwarded to the correct MAC address, or a denial of service will occur on the device you're snooping. In the case of VoIP, your PC can record (or play in real time) the media stream before passing the packets to the actual intended receiver. A classic man-in-the-middle hack, this technique is simplified by the outstanding network tool, Cain & Abel.

To get started, use Cain & Abel's host-discovery tool. Click the Sniffer tab and then the Hosts tab at the bottom of the GUI. Then, click the + in the toolbar. This will pop open a dialog where you can tell Cain & Abel to discover all of the devices on your network. I used it to discover the IP and MAC addresses of my Cisco 7960 phone (10.1.1.104) and my Asterisk server (10.1.1.10). Both are listed in Figure 6-16.

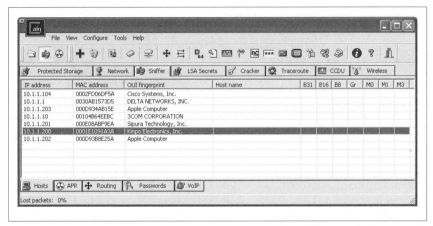

Figure 6-16. Cain & Abel's list of discovered hosts on the LAN

You're discovering all of these hosts and resolving their actual MAC addresses because Cain & Abel will need these details to perform the ARP poisoning. Once you can see the host you want to monitor in the list, click the ARP tab and then click the + toolbar icon again. You'll see the New ARP Poison Routing dialog, as shown in Figure 6-17.

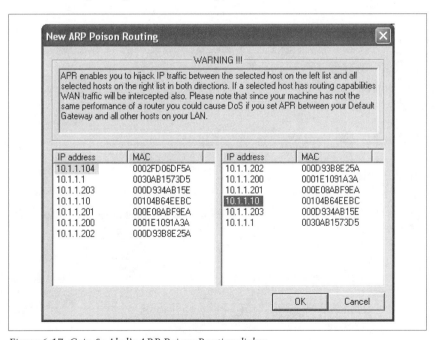

Figure 6-17. Cain & Abel's ARP Poison Routing dialog

Select the host you want to snoop on the left, and then select one or more of the destinations related to that host on the right. This way, you'll be poisoning ARP requests for the snooped hosts only if they're sent from certain IP addresses. I don't recommend that you ARP poison a large group—this could cause difficulties on the network, including a rather comprehensive denial of service—so stick to one IP address on the left and one on the right until you've got the hang of it.

Ideally, the address on the right will be the host on the other end of the call, but it doesn't have to be. If the VoIP call is to a user on the Internet somewhere (via the default gateway), you would choose the IP address of your Internet router on the right side. In fact, if you wanted to sniff your Vonage calls, you would pick the address of your Vonage ATA on the left, and the address of your Internet router on the right. Once you've got a pair of addresses selected for monitoring, click OK.

Now, start the sniffer and the ARP poison router by clicking the Start/Stop Sniffer icon and the Start/Stop APR icon (which looks like a radiation symbol). Wait for a VoIP call to be placed on the targeted host—or place one yourself on that host—and watch the call list in the VoIP tab. In a moment, an entry will pop up, indicating that the call is in progress and is being recorded into a WAV file by Cain & Abel, as shown in Figure 6-18.

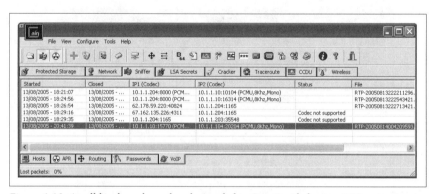

Figure 6-18. A call has been logged and recorded in Cain & Abel using ARP poisoning to intercept it

Pretty amazing, eh? One more thing: I can't emphasize enough how unethical this hack would be if performed in an environment where you aren't authorized to take these actions. I strongly urge you to be responsible with this type of hack and keep yourself out of trouble while you learn about VoIP security. Unless you have administrative authorization and the express right to monitor users' calls on the network where you're working, you should keep ARP poisoning confined to the test lab.

Hard-Core Voice
Hacks 88–100

If you've been involved in Linux VoIP hacking for very long (hey, you made it to Chapter 7, so you've been around a while), you're probably already quite familiar with Asterisk, the open source PBX (and the predominant open source VoIP platform). Aside from Asterisk, tons of open source voice networking projects are out there, including OpenH323, GnuGK, sipX, SIP Express Router, OhPhone, SaRP, and GnoPhone. Try Googling some of these. You'll find there's enough open source VoIP stuff to keep you busy for a while.

New VoIP projects arrive weekly at SourceForge.net, so this chapter represents only a partial smattering of what's out there. There's no question that the open source world is a voice hacker's paradise, a realm of mission-critical, high-stakes, real-time applications with very little tolerance for underperformance, and SourceForge is crawling with new ways to take advantage of real-time, converged networks for mission-critical voice apps.

Being a voice hacker is sort of like being a network marine. You've got to train hard to spot issues that don't show up in other, less loss-sensitive kinds of networking. You've got to be "first to fight" when problems occur on a voice network, because voice users will pick up on network slowdowns and outages before plain-old data users. It takes thick skin and quick thinking to join the ranks of the voice hacker.

I've saved some of the coolest and toughest "hard-core hacks" for the last chapter because I wanted to ease you into them. You'll need a pretty good understanding of the Session Initiation Protocol (SIP), Asterisk, and Linux to get through this chapter unscathed. So, if you haven't read the first six chapters, it would be a good idea to do so now. Have some fun with those hacks, and when you're ready to get serious, come on back.

Do you want to build and harden the ultimate PC-based telephone server? Do you want to master the old VoIP standard, H.323? How about building a fax-to-email gateway? How about building a PBX server with no hard disks, or bridging a SIP network with the Skype network?

I'm willing to bet you'll rise to the challenge.

Build a Killer Telephony Server

#88 Using any old PC with Linux is great way to experiment and learn about Internet Protocol (IP)-based telephony, but to implement a production server, you'll need some slightly bigger iron, and you'll need it hardened.

In my travels as a networking consultant, I get to visit a lot of enterprise data centers. These range from meager, stuffy, 100-degree closets crammed with desktop PCs that accidentally became servers, surrounded by a spaghetti pile of crummy cabling, to the 2,000-square-foot, raised-floor, uncomfortably cool server rooms with halon fire-prevention systems and row after row of racks filled with quad-processor servers.

When I have an opportunity to work in a modern, decked-out environment, I'm thankful that I'm not crammed in an undersized, overheated closet searching in vain for a free port on an incorrectly labeled Ethernet switch where I can plug in my PowerBook. It never ceases to amaze me just how little some folks seem to care about the environmental state of their critical data and communications equipment. As long as things keep running, I suppose they aren't likely to balk at the sorry state of their servers.

But with a critical application like telephony, from which humans have come to expect 100% reliability over many decades, a dilapidated hodge-podge of PC equipment sitting under a leaky roof just isn't going to work. A desktop-gone-server isn't going to cut it either.

Fortunately, you're about to find out how to build a voice server *right*. This means selecting the right PC or commercial telephone PBX chassis, connecting it to the right components, choosing the right operating system, and hardening it. And that means following a principled philosophy of stability, high availability, and compatibility.

The Three Things That Matter Most in Telephony

When building a server for an enterprise telephone system, you should keep three areas of focus in mind. Here they are, in my order of importance:

Stability
    The predictable, reliable operation of the server. Downtime should be nonexistent, and responsiveness should be instant.

High availability

The server (and network) must be adequate to host real-time applications without a noticeable impact when server resources are shared among many users. The server must also be able to survive a hard disk or power-supply failure without interruption to the hosted application.

Compatibility

The server, OS, and installed voice services must support well-known standards to be adaptable to changes in the business, such as growth, strategic partnerships, and geographic moves.

Creating stability. To provide the utmost stability, a server's environment must be kept cool, with a room temperature of 65 to 75 degrees Fahrenheit. It should be kept dry (duh) and connected to a protected power source with a battery backup. Large environments should consider a backup power generator with an automatic transfer switch. The switches and routers that provide connectivity to the server should be on protected power, too.

The server should also be well hardened against potential denial-of-service attacks, which you'll do shortly.

Creating high availability. To assure that the server is always available and that its voice services aren't affected by internal hardware failures like failed disk drives and power supplies, you should design redundancy into each server. A RAID 5 disk array with four or more hard disks provides a "hot spare" so that you can swap out a failed drive without having to shut down the server (you can also consider building a server with no hard disks [Hack #95]). Redundant power supplies allow the same hot-swap ability in case a power supply bites the dust. There are other techniques for high-availability, too, like clustering.

Building in compatibility. If you want to future-proof your voice server, don't bother building it on software that doesn't support the open standards needed to make it interoperable with other servers. Specifically, your software needs to support SIP and Real-time Transport Protocol (RTP). This means sticking with a highly compatible VoIP platform like Asterisk or sipX, or carefully evaluating a commercial solution.

Size and Select a Voice Server

If you've chosen a commercial softPBX like the Avaya Media Server or Cisco CallManager, you're pretty much pinned to the sizing guidelines provided by the manufacturer. When you build it yourself—on Linux or BSD—you've got total control over scalability.

With that said, there's really no hard and fast rule for determining how much processor power your server needs. I've run small workgroup (5- to 10-user) Asterisk servers on Pentium II machines. A large workgroup with hundreds of phones connected would certainly need a much beefier computer.

VoIP can be a very light load on a PC, or an immense one. If you have five SIP phones connected to a single Asterisk server, all using the simplest codec (uLaw), a Linux server with a slower processor (Pentium II or newer) and 256 MB of RAM is probably just fine. But if you have 50 SIP phones using three or four different codecs and attaching to the Public Switched Telephone Network (PSTN), you'll need at least a 2 GHz Pentium 4 or equivalent with 1 GB of RAM.

SIP-to-SIP calls are much less processor-intensive than SIP-to-Zaptel calls. Conference calls are more processor-intensive than normal two-party calls, and bandwidth-preserving codecs like the G.729 are the biggest processor hogs of all. If you have a great need to support highly compressed codecs on a lot of phones, and you'll be attaching your Asterisk server to the PSTN or to legacy phones using Zaptel, you'll need more speed, more RAM, and then more speed again.

To relieve the burden on a softPBX, try offloading processor-intensive tasks to dedicated hardware. Try to maintain an all-SIP, all-uLaw environment on your softPBX. Let off-board equipment like SIP-to-PSTN gateways (such as the Clipcomm [Hack #43]) handle codec-processing tasks to preserve capacity on the softPBX.

Select an OS and Harden It

If you're building a Linux voice server, there's not much point using an older Linux distribution. So get a recent revision of Fedora Core, burn it to CD-ROMs (four of them), and install it, keeping a few things in mind about all of those optional software packages the installer will prompt you about:

- The more optional packages you install, the less room you'll have for things like voicemail and logfiles, both of which are important in the world of telephony, right?

- The more optional packages you install (like the X Window System), the more security risks you take. Security is a good thing, right?

- The more optional packages you install, the less dedicated processing power your server will have for telephony purposes. (Are you starting to see a pattern here?)

Once Fedora Core is up and running, you're ready to start hardening. Though I've geared this discussion toward an Asterisk server, it is principally accurate for any softPBX.

 Why not use Windows for a phone server? Well, as it turns out, Windows doesn't support nearly the selection of free, high-quality telephony software such as Asterisk, sipX, OpenH323, and other open source, server-side stuff. It also turns out that Windows itself comes towing a rather pricey license fee. So I figured I'd stick to what you can download from the Net without breaking the bank. The less money you spend on licensing fees, the more you can spend on other parts of your telephony solution.

When hardening a server, you need to examine two basic aspects of the soft-PBX: the software that's installed and the software that's running. In terms of hardening the software that's installed but not running or not *needed*, the course of action is quite simple: remove it.

Remove unnecessary software. That means getting rid of Bind if you're not using it, removing Apache if it isn't needed, and unloading MySQL if you've no need for it. Just because the software isn't running doesn't mean you can't use it to facilitate a sophisticated security exploit, so remove it altogether if you don't absolutely need it.

Since we're dealing with Asterisk, you can disable a number of modules to reduce the risk of security exploits. Asterisk provides modules for all kinds of signaling protocols and telephony applications, and you might not need them all. Use the noload directive in */etc/modules.conf* to specify those that you'd like to disable:

```
noload => pbx_kdeconsole.so
noload => chan_modem.so
```

In this case, the two modules being disabled are the KDE log console module, which provides a graphical console for the KDE desktop environment, and the modem module, which is used for ISDN connectivity with Asterisk. Keeping unnecessary modules from loading also conserves memory on the server.

Clean up xinetd. *xinetd* is Fedora's catchall daemon for Telnet, finger, and a number of other Unix network applications. (It's the successor to *inetd*.) Its configuration files, in */etc/xinetd.d*, are used to enable or disable support for a long list of network-access services. Use this configuration directory to disable all but those that you absolutely need. Here's the contents of a file in this directory, */etc/xinetd.d/imap*, controlling the *imap* daemon:

```
service imap
{
        disable = yes
        socket_type                = stream
        wait                       = no
        user                       = root
        server                     = /usr/sbin/imapd
        log_on_success  += HOST DURATION
        log_on_failure  += HOST
}
```

This particular service is disabled, per the disable = yes line. Check all the files in this folder for the disable = yes line, or, if you prefer, you can altogether remove the config files for the services you don't need.

The idea is to eliminate unnecessary Transmission Control Protocol/Internet Protocol (TCP/IP) listeners, reducing the likelihood of an attacker discovering vulnerability. So, if you don't need Telnet, the Trivial File Transfer Protocol (TFTP), talk, and finger, for goodness sake, disable them. The fewer services that have listening ports, the more secure your server will be.

Optimize the local firewall on the softPBX. To build a local firewall policy on the softPBX server, you'll need to identify which VoIP protocols you're using and plan a policy based on the kind of TCP and User Datagram Protocol (UDP) port access needed by each one. Table 7-1 lists the important protocols and their respective ports.

Table 7-1. Ports used for each VoIP protocol

Protocol	Ports
SIP	5060 (TCP) and 5061 (TCP and UDP)
H.323	Both TCP and UDP on ports 2099 and 2517
H.332 Video (a.k.a. H.263)	TCP and UDP on port 2979
MEGACO/H.248	TCP and UDP on ports 2944 and 2945
TFTP	TCP and UDP on port 69
ASTMAN	5038 (TCP)
RTP	Depends on configuration of capabilities negotiation preferences of the endpoint RTP implementation; many RTP agents use 5000/5001, 5004/5005, 10000/10001, 8000/8001, or high-numbered ports
IAX	5036 (UDP)
RSVP	TCP and UDP 3455
RTSP	TCP and UDP on ports 1756, 1757, 4056, and 4057 (RTSP can vary by session like RTP)

So, if you are using SIP, you need to permit inbound SIP signaling on UDP ports 5060 and 5061.

Consider the following *iptables* policy commands:

```
iptables -P INPUT -j DROP
iptables -A INPUT -p UDP --dport 5060-5061 -j ACCEPT
iptables -A INPUT -p UDP --dport 5036 -j ACCEPT
iptables -A INPUT -p UDP --dport 5004 -j ACCEPT
iptables -P OUTPUT -j ACCEPT
```

This set of *iptables* commands manipulates the kernel's firewall such that the server can accept only RTP, Inter-Asterisk Exchange (IAX), and SIP traffic; all outbound traffic (OUTPUT chain) is permitted. This policy is based only on UDP port numbers. If incoming traffic isn't on ports 5060, 5061, 5036, or 5004, it is dropped. A truly hardened server would restrict outbound traffic, too.

For more information about securing VoIP, refer to *Switching to VoIP* (O'Reilly).

HACK #89   Build an H.323 Gatekeeper Using OpenH323

H.323 is a VoIP signaling protocol that predates SIP by five or six years, but its use in commercial telephony and desktop conferencing apps (NetMeeting, for instance) is widespread.

OpenH323 is an open source implementation of the H.323 signaling protocol suite, managed by Quicknet Technologies, the same company that makes the Internet Phone Jack line of analog interface cards. OpenH323 is distributed in binary and source code forms for both Linux and Windows, though a crafty hacker should be able to get it running on a BSD-ish OS, too.

This project will allow a Microsoft NetMeeting H.323 softphone and an OpenH323 OhPhone softphone to place calls through an H.323 gatekeeper running on a Linux computer. In this example, I'll use Microsoft NetMeeting on Windows XP and OhPhone on Mac OS X.

Although OpenH323 provides a framework of tools for developing H.323 servers and endpoints, it also natively implements a complete H.323 gateway, MCU, and endpoint. Here's a partial list of software packages that accompany OpenH323:

OpenGK
   A simple H.323 gatekeeper server example

OhPhone
   An H.323 softphone for Linux and Windows (OhPhoneX is the Macintosh version)

OpenMCU
An H.323 conference bridge server

PSTNgw
An H.323 gateway server

Each requires the base distributions of OpenH323 and its prerequisite, *PWLib*, a project-specific class library.

Installing OpenH323

A Pentium III clocked at 600 MHz should be sufficient to handle the role of a small-scale H.323 gatekeeper. The PC should be running Linux (though H.323 is also Windows compatible) and can be the same PC that runs Asterisk, if you like.

The best place to get OpenH323 is from its maintainer's web site, *http://www.openh323.org/code.html*. Compiling all of these elements is straightforward on Linux. (If you want to run OpenH323 on Windows, use the precompiled executables. The instructions I'm providing are for Linux.)

First, download and install *PWLib*. Save *pwlib_1.5.2.tar.gz* (or the filename appropriate for the version you download) to */root* as the root user. Then, unzip and untar it:

```
# tar xvzf pwlib_1.5.2.tar.gz
```

Now, you'll need to set some environment variables so that the OpenH323 software knows where to find the *PWLib* libraries:

```
# PWLIBDIR=$HOME/pwlib
# export PWLIBDIR
# OPENH323DIR=$HOME/openh323
# export OPENH323DIR
# LD_LIBRARY_PATH=$PWLIBDIR/lib:$OPENH323DIR/lib
# export LD_LIBRARY_PAT
```

If you plan to make this H.323 setup permanent, you should add the preceding environment variable commands to *.bash_profile* in */root*. Now, build the *PWLib* distribution by using make:

```
# cd $PWLIBDIR
# ./configure
# make opt
# make install
```

Next, download the main OpenH323 file to */root*. Then, unzip and untar it, substituting the filename that's appropriate for the version you download:

```
# tar xvzf openh323_1.12.2.tar.gz
```

Now, build OpenH323:

```
# cd $OPENH323DIR
# ./configure
# make opt
# make install
```

The developers recommend a 128 MB swap partition to complete the build error-free. This need is minimized if you have enough physical RAM; 256 MB of physical RAM should be plenty. This build could run for 30 minutes or more, so enjoy a delicious beverage.

Set Up the GNU Gatekeeper

Once the OpenH323 build is finished, you need to download and compile the OpenH323 Gatekeeper (*gnugk*) software. Don't confuse this with the *opengk* software that comes as a part of the OpenH323 distribution. *This* gatekeeper comes from a different source altogether but is built using the same libraries as *opengk*. The big difference is that *gnugk* is a much more complete implementation of a gatekeeper, and *opengk* is a reference example and is not very useful yet.

First, download and save the *gnugk* source code from *http://www.gnugk.org/h323download.html* into */root*. It will be named *gnugk-2.0.8.tgz* or something similar. After the download is finished, build the *gnugk* package:

```
# tar xvzf gnugk-2.0.8.tgz
# cd openh323gk
# make opt
```

Now, issuing the *gnugk* command will launch the *gnugk* gatekeeper. If you receive an error indicating shared libraries cannot be located, make sure you've got those environment path variables set in your login profile. If you run into compiler errors, try grabbing the x86 Linux executable from the *gnugk* site. Regardless of whether you compile it yourself, copy the contents of the package's *bin* directory into */usr/sbin* and the contents of its *etc* directory into */etc*, as follows:

```
# cd openh323gk
# cp bin/* /usr/sbin
# cp etc/gnugk.ini /etc
```

To install a sample config file that allows any endpoint to register with the gatekeeper, copy *etc/proxy.ini* also:

```
# cp etc/proxy.ini /etc
```

proxy.ini is far more permissive than the default configuration file, and it will allow you to register unauthenticated (i.e., passwordless) endpoints. Now, you can run *gnugk* with the config file in */etc* by issuing the following:

```
# gnugk -c /etc/gnugk.ini
```

Register an H.323 Softphone Using OhPhoneX

If you're using a Windows PC, chances are you already have Microsoft Net-Meeting. This is a very capable softphone, and it works well with OpenH323. In fact, the next section describes how to set it up.

But since the OpenH323 project produces a phone, too, we'll use it. That phone is called OhPhone, and it's distributed as an executable for Linux, Windows (*http://www.openh323.org/*), and Mac (*http://xmeeting.sourceforge.net/*).

These examples use screen grabs from the Mac OS X version. The Linux and Windows versions have only a text-based UI, but for those platforms, GnomeMeeting and MS NetMeeting make great alternatives.

The first thing you'll need to do with OhPhoneX is access its Preferences menu option. The Gatekeeper tab of the Preferences window will allow you to specify a gatekeeper, username, password, alias, and E.164 address (phone number), as shown in Figure 7-1.

In Figure 7-1, the address of the gatekeeper is 10.1.1.10. The ID is a superficial, freeform ID used like caller ID. The User Alias/ID is required only if *gnugk* is configured for authenticating registration attempts. The password field is optional; its use is policy-dependent, as *gnugk* accepts blank passwords. Finally, the E.164 number is the phone number to which the endpoint is registering and, ultimately, the phone number that will be used to route calls to this softphone. Be sure to check the "Use gatekeeper" checkbox, too.

When you close the Preferences window, click the Start Phone button and then click the Console button. You'll see whether the softphone's registration attempt with the H.323 gatekeeper was successful. The console log of OhPhoneX, shown in Figure 7-2, contains the details of the registration attempt.

Now, if you register a second softphone from a second PC, you can call back and forth between them using the gatekeeper as the E.164 alias translator. This works the same way with H.323 hardphones. Callers dial the E.164 digits, and the gatekeeper provides the E.164 "resolution" that allows the software in the phone to do its H.225, H.245, and RTP signaling to facilitate the call.

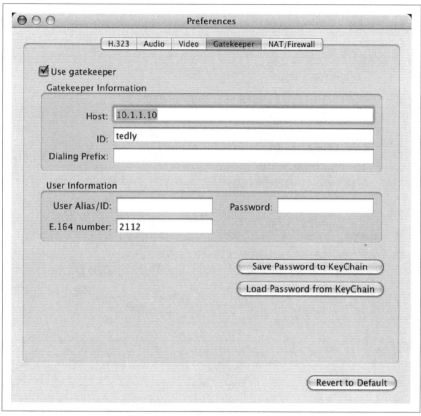

Figure 7-1. OhPhoneX's Preferences window has all the options an H.323 endpoint could possibly need to register with a gatekeeper

Once a call is in progress, the Connection Stats window shows the status of the call in excellent detail, as in Figure 7-3.

Register an H.323 Endpoint Using NetMeeting

Microsoft NetMeeting is an H.323 softphone application that comes packaged with Windows Me, 2000, and XP. To run it on XP, however, you'll have to perform a slight hack to activate it. Select Start Menu → Run, type **conf**, and click OK. Then, select "Put a shortcut to NetMeeting on my desktop" in the wizard that follows. Once this is done, NetMeeting is activated on Windows XP just as it would normally be on Windows 2000.

To configure NetMeeting to register with the gatekeeper, inside NetMeeting click Tools → Options. This will display the Options dialog, where you can click the Advanced Calling button. The Advanced Calling Options dialog box will appear, as in Figure 7-4. Check the boxes next to "Use a gatekeeper

Figure 7-2. OhPhoneX's console log can help you troubleshoot the registration process

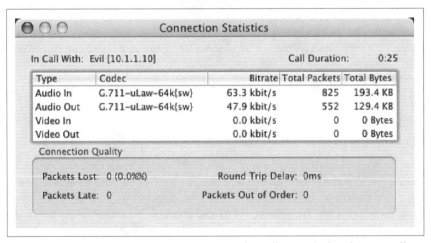

Figure 7-3. OhPhoneX's Connection Statistics window tells you which codec your call has selected and how much bandwidth it's using

to place calls" and "Log on using my phone number." In the "Phone number" field, enter the address of the gatekeeper, as well as the E.164 address you'd like to use.

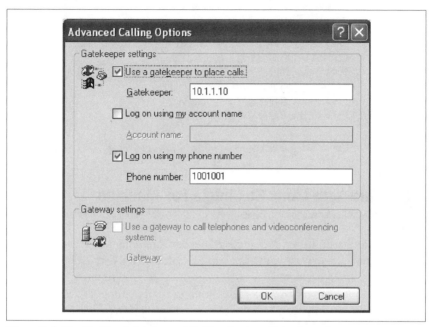

Figure 7-4. The NetMeeting Advanced Calling Options dialog box allows you to configure gatekeeper registration

Microsoft NetMeeting is a very worthwhile H.323 softphone, and it's quite customizable. It allows video calling as well as audio calling, and it has a built-in T.120 whiteboard and instant-messaging (text chat) applications. You can tweak the codec-selection preferences by choosing Audio from the Options dialog and then clicking Advanced. The codec-selection dialog is shown in Figure 7-5. If you're really looking to restrict codec selection, most compliant gatekeepers allow you to do it centrally.

Make the Call

Once both phones are registered with the gatekeeper, you can call between them using their E.164 numbers, since they're on the same zone. Now, if you like, download OpenAM from the OpenH323 project to set up an H.323-based personal message recorder.

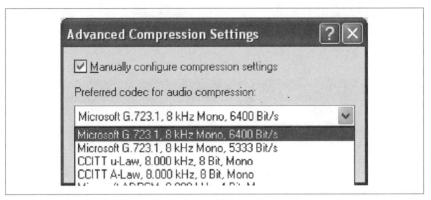

Figure 7-5. NetMeeting ships with a selection of five codecs, including G.711 (uLaw/aLaw) and G.726 (ADPCM)

HACK #90 Turn Your Linux Box into a Fax Machine

Have you ever wished you could handle fax traffic with your Linux machine?

Asterisk offers a built-in fax-detection mechanism. This allows you to handle faxes that are sent to your Asterisk box on a Plain Old Telephone Service (POTS) line connected via a Zaptel interface. It's Asterisk's Answer( ) command that triggers the fax detection. If an incoming fax is detected, Asterisk automatically transfers the call to the special extension called *fax*, if it exists.

To use this special extension, you'll need to compile and install the *spandsp* package. Download the latest version from *ftp://ftp.opencall.org*, and unzip the file into */usr/src/spandsp*. To compile it, issue these commands:

```
# ./configure --prefix=/usr
# make; make install
# cp app_rxfax.c /usr/src/asterisk/apps
# cp app_txfax.c /usr/src/asterisk/apps
# cp Makefile.patch /usr/src/asterisk/apps
# cd /usr/src/asterisk/apps
# patch <Makefile.patch
```

> If you're worried about the security concerns associated with compiling as root, you can use a nonroot account to compile *spandsp*.

These commands compile the *spandsp* package, which provides a source code patch for Asterisk. As such, you'll need to recompile Asterisk now:

```
# cd /pathtoasterisksource/asterisk
# make clean ; make install
```

The next step to faxing with Asterisk is to enable fax detection on the Zaptel channel you want to use for faxing. This doesn't stop the channel from being used for normal voice calls; it just enables the channel to discern fax calls from normal calls. To enable this function, be sure that the channel's section in */etc/asterisk/zapata.conf* has this entry:

```
faxdetect=both
```

The valid parameters for the faxdetect option are incoming, outgoing, both, and no. By default, fax detection is disabled.

Receiving Faxes

Now, consider the following snippet from a dial plan:

```
[incoming-local]
exten => s,1,Answer
exten => s,2,Dial(SIP/202,45,rm)
exten => s,3,Voicemail(202)

exten => fax,1,SetVar(TIFFILE=/var/spool/faxes/thisfax.tif)
exten => fax,2,rxfax(${TIFFILE})
```

In this context, the Answer() command triggers fax detection. If the call isn't a fax, the dial plan calls for a call to SIP peer 202. If it is a fax, the fax extension takes over, saving the fax image into a TIFF file located in */var/spool/ faxes*. Another script can then process the file in any way you see fit, perhaps printing it immediately, like this:

```
exten => fax,1,SetVar(TIFFILE=/var/spool/faxes/thisfax.tif)
exten => fax,2,rxfax(${TIFFILE})
; dump the FAX file to the default printer and remove the FAX file
exten => fax,3,System('tiff2ps ${TIFFILE} | lpr')
exten => fax,4,System('rm ${TIFFILE}')
```

 tiff2PS is a utility provided in the *libtiff* package, a library for dealing with TIFF files. It's a standard part of many Linux distributions, Red Hat included.

Sending Faxes

Receiving faxes with Asterisk is quite a bit easier than sending them, because when receiving them, the work of scanning them into digital form is done already. This is the part neither Asterisk nor *spandsp* addresses. However, these packages *can* very easily fax a TIFF file. So it's up to you to get that TIFF file in a path where *spandsp* can grab it.

This can happen in any number of ways. You can create a simple web interface that allows you to upload TIFF files to the server, or, if you have the right software, you can just scan them directly using the Linux machine. I don't recommend either of these approaches, however, because neither of them provides a straightforward way of telling Asterisk where to send the fax from the outside application or script that's handling the scanning and packaging.

Without a lot of hacking, Asterisk just doesn't make a good day-to-day, occasional-use fax server for outbound fax transmittals. There are better solutions to this need already. One of them is HylaFAX, a freely available fax server for Linux and BSD operating systems. HylaFAX can use standard fax/modems, which also makes it cheaper to implement than Asterisk with (comparatively expensive) Digium voice cards. You can obtain HylaFAX from *http://www.hylafax.org/*.

HACK #91   Build an Inbound Fax-to-Email Gateway

Once you start faxing with your Linux box, why stop there? This hack shows you how to route faxes automatically into emails and PDF files.

In the previous hack, you built a configuration to direct all incoming faxes from Zaptel channels to a file, which, in turn, you could automatically print. But, if the server were working on behalf of many possible fax recipients, you would have to rely on the incoming fax's cover sheet to know which recipient it's destined for. Worse still, someone would have to go to the printer, pick up the fax, and hand-deliver it to the correct person.

There's a better way, of course: email. It's just as easy to email that TIFF file to somebody's inbox as it is to print it. Here is a dialing plan that will do just that:

```
exten => fax,1,SetVar(TIFFILE=/var/spool/faxes/thisfax=${CALLERIDNUM}.tif)
exten => fax,2,rxfax(${TIFFILE})
; email the FAX file to the receptionist and then delete it
exten => fax,3,SetVar(EMAIL=receptionist@oreilly.com)
exten => fax,3,System('mewencode -e ${TIFFILE} | mail -s FAX ${EMAIL}')
exten => fax,4,System('rm ${TIFFILE}')
```

This configuration receives the fax, MIME-encodes it using the `mewencode` command (a standard part of most Linux distributions), and emails it to the email address stored in the `${EMAIL}` variable. This is a catchall solution; it sends every fax that's received to the same recipient, who can then forward it (and screen it if necessary) to the appropriate person based on the content of each fax message.

Automatic Fax Routing

To have the Linux server automatically route each fax to the right recipient (instead of having a certain email user doing it), we must have a way of associating each fax with the correct recipient. We'll have to associate a certain line (or a certain DID) with each user so that whenever a fax is received on that line (or DID), we'll know where to route it. Each phone line's number, or each DID number, if we're using a primary rate interface (PRI), will become a single user's fax number.

To associate a DID with a particular user's email address, we can use an LDAP inquiry. An LDAP client library for Linux, *openldap*, provides applications with the ability to access LDAP servers and perform such inquiries. Your LDAP server might be an Exchange or Lotus Domino server where directory information is stored. The Red Hat distribution includes OpenLDAP binary package and the OpenLDAP developer package.

But you'll need more than just the LDAP client library. You'll also need an actual LDAP client for Asterisk, such as Sven Slezak's LDAPget package. Download it from *http://www.mezzo.net/asterisk/* and unzip it into the Asterisk source directory, */usr/src/asterisk*.

Next, as root, copy the *app_ldap.so* file into */usr/src/asterisk/apps*. Then, use a text editor to add *app_ldap.so* to the list of applications that begins with APPS= in */usr/src/asterisk/apps/Makefile*. While you have the *Makefile* open, add the following rule just above the *app_voicemail.so* line:

```
app_ldap.so : app_ldap.o
    $(CC) $(SOLINK) -o $@ $< -llber -lldap
```

Now, LDAPget is ready to be compiled. Save your changes in the text editor and exit back to the shell, where you'll issue these commands to compile the package:

```
# cd /usr/src/asterisk
# make; make install
```

If Asterisk isn't currently running, start it. Then, go to the Asterisk command line and load the LDAPget module (alternatively, you can just restart Asterisk):

```
pbx*CLI> load app_ldap.so
pbx*CLI> show application LDAPget
```

The show application command will confirm that the module is installed and loaded by showing you a brief description of the LDAPget dial-plan command. Now, you can set up the LDAP inquiry your dial plan will use to get email addresses based on the DID provided by ${EXTEN}. To set up this query, open */etc/asterisk/ldap.conf*. It might not exist yet, since you've only just compiled the LDAP module. Create an entry like this in *ldap.conf*:

```
[mailfromdid]
host = ldap.oreilly.com
user = cn=root,ou=People,o=oreilly.com
```

```
pass = jarsflood
base = ou=Addressbook,o=oreilly.com
filter = (&(objectClass=person)(|(fax=%s)))
attribute = email
convert = UTF-8,ISO-8859-1
```

This configuration will cause an LDAP inquiry to ldap.oreilly.com, asking for an object of the person class with the attribute fax equal to the value of the %s token (which will be replaced with the DID at runtime). The attribute setting tells the inquiry which attribute from the object to return as a value to the dial plan's variable. This might seem confusing right now, but it should be clearer once you see how the LDAPget command is used in the dial plan.

In the context where your incoming PSTN calls begin (specified in *zapata.conf*), you can capture the DID from the ${EXTEN} variable and use it to supply an argument to an LDAP inquiry. If the inquiry is successful, Asterisk's LDAP client will return the email address with which the fax number (i.e., DID) is associated, as in this snippet of *extensions.conf*:

```
[incoming-pstn]
exten => s,1,SetVar(DID=${EXTEN})
exten => s,2,Answer
exten => s,3,Ringing
; allow 4 seconds for the FAX detection
exten => s,4,Wait(4)
; if no FAX, send this call to be handled elsewhere
exten => s,3,GoTo(incoming-voice)

; here's the fax handling extension, which sends the call to the
; 'inc-fax' context
exten => fax,1,Goto(inc-fax,1,1)

[inc-fax]
exten => s,1,SetVar(TIFFILE=/var/spool/faxes/${DID}.tif)
; The 'mailfromdid' LDAP inquiry is defined in Asterisk's ldap.conf file.
exten => s,2,LDAPGet(EMAIL=mailfromdid/${DID})
; If the LDAP inquiry succeeds, priority will be 2+1.
exten => s,3,rxfax(${TIFFILE})
exten => s,4,GoTo(105)
; If the LDAP lookup fails, priority will be 2+101.
exten => s,103,SetVar(EMAIL=receptionist@oreilly.com)
exten => s,104,rxfax(${TIFFILE})
; Now, e-mail the FAX file to whichever e-mail address was decided upon.
exten => s,105,System('uuencode ${TIFFILE} uuenc | mail -s FAX ${EMAIL}')
exten => s,106,System('rm ${TIFFILE}')

[incoming-voice]
; non-fax calls are handled here
```

The result of all of this compiling and config tuning is that different email recipients now have assigned fax numbers on the PRI (or assigned POTS lines for their exclusive use as inbound fax lines). When you send a fax to Todd's fax number, Todd receives the email. When you send it to Susie's, Susie receives the email, and so on. Of course, it's up to you to populate your LDAP server with the right information and to make sure the inquiry config in *ldap.conf* matches your LDAP server's schema.

> Don't have an LDAP server? You can use Asterisk's built-in database commands to resolve DIDs to email addresses. Chapter 17 of *Switching to VoIP* (O'Reilly) contains a command reference that covers these dial-plan commands.

Hacking the Hack

In the previous hack, you used the `tiff2ps` command to create a printable version of the fax, but with a few extra steps, you can turn a TIFF into a PDF file, too. PDF can be preferable to TIFF when using email, as we are in this project. Consider the following dial-plan changes to the [`inc-fax`] context:

```
exten => s,105,System('tiff2ps -2eaz -w 8.5 -h 11 ${TIFFILE}| ps2pdf \
>${TIFFILE}.ps')
exten => s,106,System('uuencode ${TIFFILE}.ps uuenc | mail -s FAX ${EMAIL}')
exten => s,107,System('rm -f ${TIFFILE}*')
```

Now, instead of just encoding the TIFF file and emailing it, the file is converted to a PostScript file and then to a PDF file, before being uuencoded and emailed to the appropriate recipient.

Teach Your Asterisk Box to Speak
HACK #92

Sometimes you just don't want to get off the couch and walk to the caller ID display. Your Linux server understands and wants to help.

My first exposure to synthesized speech was on a Commodore 64; the speech demo took an eternity to load off a floppy diskette, and the speech sounded like an English as a Second Language student was speaking it directly into a pillow. Today, with DSP and decades of additional speech-programming research in the bag, synthetic speech is much more passable, and folks are constantly coming up with novel uses for it.

In Detroit, I have a buddy whose Linux server used to announce logfile entries and tell him when the doors around his house were opened and closed. While this speaking server was mysteriously silenced right around the time he got married, I still love his hack. Adapting speech capability around Asterisk is a logical use for two of my favorite pieces of open source software: Asterisk and Festival, the University of Edinburgh's speech synthe-sizer. With a little bit of dial-plan configuration, your Linux Asterisk server will be announcing your incoming calls in no time (and announcing a whole bunch of other stuff, if you want).

> Mac OS X users will find that Festival is similar in some ways to the Say command on OS X, though Festival pro-vides much more functionality to Asterisk than Say does. For example, you cannot use Say to speak to callers, as you can with Festival.

First, you'll need a sound card installed and working in your Linux box. Most commercial distributions of Linux (Red Hat, Debian, etc.) make sound card configuration a straightforward affair. Next, get your hands on the Fes-tival source code at *http://www.cstr.ed.ac.uk/projects/festival/download.html*. Find a link to download it, and save it from your browser, or use wget to grab it (and its supporting Speech Tools libraries):

```
# wget http://www.cstr.ed.ac.uk/downloads/festival/1.95/festival-1.95-beta.
tar.gz
# wget http://www.cstr.ed.ac.uk/downloads/festival/1.95/speech_tools-
1.2.95beta.tar.gz
# wget http://www.cstr.ed.ac.uk/downloads/festival/1.95/festlex_CMU.tar.gz
# wget http://www.cstr.ed.ac.uk/downloads/festival/1.95/festlex_POSLEX.tar.gz
# wget http://www.cstr.ed.ac.uk/downloads/festival/1.95/festvox_kallpc_16k.
tar.gz
```

Once you've downloaded Festival, you should unpack, compile, and install it with these commands:

```
# tar xvzf speech_tools-1.2.95-beta.tar.gz
# tar xvzf festival-1.95-beta.tar.gz
# cd speech_tools
# ./configure
# make; make install
# tar xvzf festlex_CMU.tar.gz
# tar xvzf festlex_POSLEX.tar.gz
# tar xvzf festvox_kallpc16k.tar.gz
# cd festival
# patch -p1 </usr/src/asterisk/contrib/festival-1.4.3-diff
# cd ../festival
# ./configure
# make; make install
```

The make; make install commands take the longest—upward of five min-utes each on my trusty old garage-built Pentium III machine. Of course, if you've already compiled and installed Asterisk [Hack #41], installing Festival will seem fast.

While you're waiting for the compile to complete, let me give you a quick Festival crash course. Festival has an interactive mode, where you can issue speech commands, as well as an "execute and exit" mode, where you can pass instructions to it from the Unix shell. Simply executing festival in a shell will put you in interactive mode, where you can interact with the speech synthesizer:

```
festival> (SayText "Hello world.")
festival> (tts "text-file.txt")
```

The SayText command simply causes Festival to speak the quoted text using your PC's sound card, and the tts command speaks the contents of the text file indicated. By the way, if you'd like to quit interactive mode and return to the shell, hit Ctrl-D.

When you execute Festival from the command line, you've got some cool functionality at your disposal. Executing with --pipe causes Festival to take commands from standard input. Recent builds of Festival also include the *text2wave* application, which generates Wave-format sound files from text input.

The Hack

For Asterisk to support text-to-speech via Festival, you have two approaches. The first is to use Asterisk's built-in Festival() command, which is a standard part of the Asterisk distribution, when patched for Festival as described earlier. To get to this command, you might need to recompile Asterisk after the Festival patch has been applied. The Festival installation instructions provided earlier show how to apply this patch. In the Asterisk *extensions.conf* file, the Festival command simplifies the playing of synthetic speech for callers:

```
exten => s,1,Answer
exten => s,2,Festival('Hello caller. My name is Mr. Synthetic.')
```

To greet a caller by name, use Asterisk's built-in caller ID variables:

```
exten => s,1,Answer
exten => s,2,Festival('Hello ${CALLERIDNAME}. My name is Mr. Synthetic.')
```

The Festival command allows you to send speech to callers, but not to the sound card. To do that, you'll need to pump some output from Asterisk to the Festival application at the appropriate time in the dial plan. Using Asterisk's System() command, you can trigger all kinds of activity in the Unix shell from within Asterisk—including, of course, Festival activity. Take a look at this Festival shell command, which simply speaks the quoted text through the PC's sound card:

```
# echo "Hello world" | festival --tts
```

This causes the text output of the echo command to flow to Festival's standard input. It is then spoken using the audio output of the sound card. Commands like this really give Asterisk some cool speech abilities. Say you wanted to have your Asterisk server announce the caller ID of each incoming call:

```
exten => s,1,System(echo "You are receiving a call from {$CALLERIDNAME}" \
| festival --tts)
exten => s,2,Dial({$DEFAULTPHONES})
```

Mac the Hack

If your Asterisk server runs on a Macintosh [Hack #93], you can do the caller ID announcements using the Mac Unix command, Say, instead of Festival. One of the coolest things about the Mac's built-in speech synthesis is its selection of different voice styles; males, females, whispers, and hysterical laughing all make Mac speech a lot of fun. In fact, you can use the Say command's -v option to use those different voices, depending on the variables of the call. Here, different voices are used depending on which phone is being dialed:

```
exten => 10,1,System(say -v Agnes Kelly, you have an incoming call.)
exten => 10,2,Dial({$KELLYSPHONE})
exten => 20,1,System(say -v Hysterical Jake, you have an incoming call.)
exten => 20,2,Dial({$TEDSPHONE})
```

The preceding sample uses the Agnes voice to announce Kelly's calls and the Hysterical voice to announce Jake's calls.

HACK #93 Build a Mac PBX

The Mac mini is a very tiny and rugged PC, making it a great small-office PBX.

When Apple introduced the Mac mini, most onlookers were pleased to see a smaller machine with plenty of muscle—enough to handle a few dozen VoIP phone calls at a time—even though most observers didn't have VoIP in mind for it. At less than $500, the mini is great for the cost-conscious, and for those who don't trust the likes of Windows in a real-time application like IP telephony, a Mac provides a secure, friendly alternative. Of course, there are comparably equipped "small" PCs, but none with the tiny (two inches tall, and six by six inches square) form factor of the Mac mini. So, if you need a space-conscious, cheap VoIP server, you need look no further than the mini.

> Would you rather use an Xserve with this hack? Great idea! The Xserve has faster processors and a RAID hard-drive array. This means high-performance PBX action. It also has an extra Ethernet interface and a swappable power supply, making it better in mission-critical situations than the mini.

But, since the Mac mini has no card slots (and since multichannel PCI telephony drivers are not available for OS X), attaching analog phones and phone lines to a Mac mini isn't the same as in a traditional, PCI-equipped server chassis, where you can snap foreign exchange office/foreign exchange station (FXO/FXS) cards into place to connect phone lines. This is where VoIP comes in handy. Just because the Mac mini can't connect directly to analog lines and phones doesn't mean you can't use them with it.

You can connect analog lines and phones to a Mac mini by way of an analog telephone adapter (ATA) or an FXO gateway device. An ATA will connect an analog phone to an Ethernet network, and an FXO gateway will attach a phone line to the Ethernet network. These devices provide a signaling proxy that allows analog phones and lines to be used with VoIP servers like that Mac mini PBX we're about to set up. The Clipcomm CG-410 referenced elsewhere in this chapter is one such device—an FXO gateway device.

This hack works on any Mac with OS X 10.2 or higher. To get started, let's download an installation package for a Mac-compatible distribution of Asterisk. The one I like was built by Benjamin Kowarsch, a Mac telephony hacker. You can download it from *http://www.astmasters.net/* or *http://www.macvoip.com/*. You'll probably want to grab the latest version available from one of those sites. Unpack it (StuffIt should prompt you to unpack it as soon as you download it) and launch the *Asterisk.pkg* file. This package looks like most Mac installer packages. Step through the Introduction, Read Me, and License screens; agree to the license; and select the volume where you want to install Asterisk (Figure 7-6). Your boot volume is the only place you can install Asterisk, so don't bother yourself with trying to figure out how to install it elsewhere. Besides, unless you like time-consuming nondefault settings that require the constant attention of a Unix snob, the boot volume should be adequate.

The quickest way to launch Asterisk (on any system), once it's installed, is to issue this command:

```
$ sudo /usr/sbin/asterisk -vvvc
```

The more v's you tack on, the more verbose Asterisk's debugging output will be—very useful when troubleshooting things.

If you've issued this command in the OS X terminal and received the following message, or something similar, you weren't using the root account when you launched Asterisk:

Logger Warning: Unable to open log file '/var/log/asterisk/messages': Permission denied

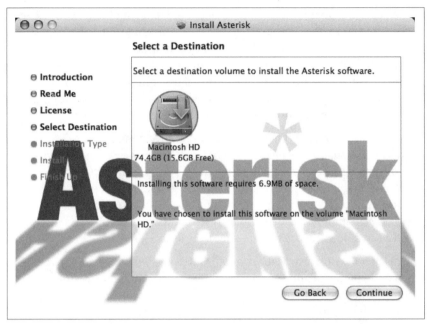

Figure 7-6. The Asterisk installer package

You'll need to be root, the all-powerful Unix administrator account, to launch Asterisk as installed by the installer package. Becoming root on OS X isn't as simple as it is on Linux, however. The root account itself is actually hidden away and disabled until you go in and "turn it on" manually. This practice is generally frowned upon, as using the root account gives you enormous capacity to harm your Mac's filesystem and settings, so don't say I didn't warn you. We're just going to do this once, so you can see how it's done to launch Asterisk, and then we'll return to the safe zone of nonroot access. As the default user (the first user created when you first installed OS X on your Mac), open a terminal and issue this command:

```
$ sudo passwd root
```

You'll be prompted for a password. This password will establish the root user's password so that you can use the root account from now on. Yes, that simple command "turns on" the root account. Now, you should be able to launch Asterisk on the Mac using the preceding Asterisk command. If you get a few screens full of log output that eventually ends up at a PBX prompt, you've launched Asterisk successfully.

Yet even after all that, Asterisk won't be set to start automatically every time you boot your Mac, so you'll need to make sure it's launching every time. This is easy. In */System/Library/StartupItems*, create a folder called *Asterisk*.

In that folder, create a text file called *Asterisk* and place the following code
inside:

```
#!/bin/sh

##
# Ted's Asterisk Telephony Server Startupitem
##

. /etc/rc.common

StartService ()
{
        echo "Starting Asterisk telephony"
        /usr/sbin/asterisk -vvvg &
}

StopService ()
{
    echo "Stopping Asterisk telephony"
    /usr/sbin/asterisk -rx "stop when convenient"
}

RestartService ()
{
    echo "Restarting Asterisk"
    StopService
    StartService
}

RunService "$1"
```

This script handles the proper startup (and shutdown) of Asterisk when the
system boots (and shuts down). Likewise, rebooting your Mac PBX will also
automatically shut down and restart Asterisk. Before you reboot, though,
you'll also need to create a file called *StartupParameters.plist* in the same
folder—a text file with the following contents:

```
{
  Description    = "Asterisk";
  Provides       = ("Asterisk");
  Uses           = ("Disks", "NFS");
}
```

Once those two files are in place, give your OS X machine a reboot, and
your Mac PBX is built. Now all you need is somebody to call.

Monitor Asterisk from Your Perl Scripts

HACK
#94

If you've used Linux (or FreeBSD or Mac OS X or Solaris) for longer than an hour, chances are good you've used Perl. Now, use Perl to monitor and control your Asterisk PBX.

The Perl module of choice for Asterisk is appropriately called *asterisk-perl*. It provides connections between the Asterisk Gateway Interface (AGI) and the venerable scripting language named after a misspelled maritime phenomenon. It also links Perl with Asterisk's Manager interface, a socket application programming interface (API) that lets you control and monitor Asterisk by sending messages to it on a TCP port—5038 to be exact.

For ad hoc interaction with the Asterisk Manager, you can telnet to that port on your Asterisk server—if the manager is enabled, that is. To ensure Asterisk Manager is indeed running and able to respond to your requests so that your Perl programs will actually *do something* once *asterisk-perl* is installed, you need to pay a visit to */etc/asterisk/manager.conf*. Make it look roughly like this, being sure to include enabled=yes and to add a section like the [hansolo] one to define a username and password with which to access the Asterisk Manager later on:

```
;
; Asterisk Call Management support
;
[general]
enabled = yes
port = 5038
bindaddr = 0.0.0.0

[hansolo]
secret = falcon
deny=0.0.0.0/0.0.0.0
permit=127.0.0.1/255.255.255.0
read = system,call,log,verbose,command,agent,user
write = system,call,log,verbose,command,agent,user
```

You'll need to restart Asterisk (run asterisk -rx as root at the shell prompt) to commit these config changes. Next, you'll need to install the Asterisk Perl module. Download, compile, and install it as follows:

```
# wget http://www.netdomination.org/mirror/asterisk.gnuinter.net/ \
files/asterisk-perl-0.08.tar.gz
# tar xvzf asterisk-perl-0.08.tar.gz
# cd asterisk-perl-0.08
# perl Makefile.PL
# make all
# make install
```

Now, pop into your *asterisk-perl* source directory and check out the Asterisk Manager example in the *examples* directory. It's called *manager-test.pl*, and it demonstrates how to poke Asterisk with Perl. For it to work, though, it will need to be authenticated as a legitimate Asterisk Manager API user, and that means adjusting the beginning of the script to match the username and password you put in */etc/asterisk/manager.conf*. Open *manager-test.pl* and make the username and password match:

```
$astman->user('hansolo');
$astman->secret('falcon');
$astman->host('localhost');
```

localhost is used to specify the host that the Perl script will connect to in order to send messages to the Asterisk server; in this case, it'll connect to the same machine as the one where the script is running. Run the script like this:

```
# ./manager-test.pl
```

If you place a call to any of the channels on the Asterisk server, the script will give you output like this, via its connection to Asterisk Manager:

```
Event: Newchannel
Uniqueid: 1121992811.1
Callerid: <unknown>
Channel: Zap/3-1
State: Ring
Event: Newexten
Channel: Zap/3-1
Context: default
1121992811.1:
Extension: s
Application: Answer
AppData: Uniqueid
Priority: 1
Event: Newstate
Callerid: "Cleveland     OH" <4403281441>
Channel: Zap/3-1
State: Up
Uniqueid: 1121992811.1
```

This output says that the Asterisk server has received a call from 440-328-1441 on channel Zap/3, assigned it a unique ID (for tracing it among the other Asterisk Manager output), and indicated that it is being handled by extension s (the default extension) in the default context. The State: Ring bit indicates that the channel is merely ringing and hasn't yet been answered. The Application: Answer line indicates that, in accordance with the dial plan in */etc/extensions.conf*, this call is being handled by the Answer() command. Quite a lot of useful information for a short, simple Perl program.

Besides viewing status output (using the `eventloop` method of the `Asterisk::Manager` class), you can also use the *asterisk-perl* Perl extensions to issue commands to Asterisk. Consider this simple Perl script:

```perl
#!/usr/bin/perl
use lib './lib' '/..lib';
use Asterisk::Manager;
my $astman = new Asterisk::Manager;
$astman->user('hansolo');
$astman->secret('falcon');
$astman->host('localhost');
$astman->connect || die $astman->error . "\n";
print STDERR $astman->command('iax show peers');
```

The output of this script, at least on my Asterisk server, which has a single permanent IAX peer set up, looks like this:

```
Name/Username Host               Mask              Port  Status
mtech/199     68.46.190.45  (S)  255.255.255.255   4569  Unmonitored
```

Of course, with the command method of the `Asterisk::Manager` class, you can send any Asterisk console command and get its output. So, you can grab the whole dial plan using `show dialplan` or `zaptel show channels` to show all the current Zaptel activity on the system. Once you get that output, you've got to parse it. Otherwise, your Perl program won't be able to do much more than print it out. So I recommend that you brush up on those great Perl text-parsing functions.

If you're familiar with the Asterisk Manager API command structure, you can also send API commands, allowing you to originate phone calls, hang up and transfer calls in progress, and do other fun things. There's a nice Asterisk Manager API reference at the end of my book *Switching to VoIP* (O'Reilly). There now—you have the tools to build an Asterisk empire using Perl and Perl alone. Go now, and conquer.

HACK #95 Build a SoftPBX with No Hard Drive

You don't really need a hard disk to run a phone system; even an IP-based softPBX doesn't need a hard drive. A CompactFlash-based PC will do the trick.

Sometime around September of 2004, I was looking at one of PC Engines' WRAP boards and wondering how well it could run Asterisk. Knowing that I would not want to run a full-size distribution, I started pulling apart a Gentoo install, removing components that are not critical to the functionality of Asterisk. After a fairly significant amount of work, I was left with a slimmed-down Gentoo that fit on a 256MB CompactFlash card (which was the smallest that I had at the time) and would run mounted, read-only. After working on the init system and writing some extra scripts, I decided to put it up on my web site just in case someone else found it interesting or useful. I decided to call it AstLinux, version 0.1.0. After about 4,000 downloads, I think that I had my answer, and AstLinux was born!

By 2005, I realized that to make AstLinux truly spectacular, I was going to have to make it smaller and more flexible. Work on AstLinux 0.2.x began. After messing around quite a bit with different build systems and methodologies, I found and stuck with a wonderful combination of crosstool and PTXdist. After some serious time and effort, AstLinux was reborn, and this time it came in at just under 27 MB—small enough to fit on a 32MB CF card.

This hack will show you how to use AstLinux to create a softPBX system that doesn't require a hard drive. Read the next section to find out the kinds of features you can fit into such a tiny system.

Current Features of AstLinux

As of this writing, AstLinux has the following features:

• DHCP server/client
• File transfer protocol (FTP) server
• TFTP server
• Asterisk (with *zaptel* and *libpri*)
• Sangoma WANRouter with voice time division multiplexing (TDM) support
• Web server with HTTPS
• Administration via console, serial console, SSH, or web graphical user interface (GUI)
• Network time protocol (NTP) client/server
• VPN support (IPSEC IKE and OpenVPN)
• SPI firewall (*iptables* with my *astfw* script)
• Quality of Service (QoS; my AstShape script)
• NFS client/server
• Linux 2.6 kernel
• Caching DNS proxy/server (*dnsmasq*)

Additionally, AstLinux now runs on everything from the Soekris net4801/ PC Engines WRAP series of Single Board Computers (SBCs) to Dell rack mount gear. Pretty much any modern machine using PC hardware is now supported by the AstLinux i586 image.

AstLinux's Keydisk

One of the more interesting concepts of AstLinux is the use of a single configuration file and the concept of a *keydisk*. In AstLinux, you can configure almost all of the system (with the exception of Asterisk itself) in one configuration file, */etc/rc.conf*. */etc/rc.conf* is a very simple text file with VARIABLE NAME = VALUE pairs. So, for instance, to set the IP address on the external interface, you would uncomment EXTIP and change it to the desired network address. You will also want to change EXTNM, etc., but I will cover that in more detail later.

Now for the keydisk. This is a perplexing concept to some people, and it can be difficult to explain. Think of it as a personality, similar to a SIM card in a Global System for Mobile (GSM) phone. The partition that AstLinux resides in is purely for AstLinux. No user files or configuration is stored there; this is how it can stay mounted read-only and how the system can still function. Also, it provides a ton of flexibility and allows for some very interesting uses of AstLinux.

When the system first boots, you will see several entries in the GRUB bootloader. They all boot AstLinux; they just pass different arguments to the kernel that the startup scripts then look at to determine what to do. One of these arguments is astkd=. So, for the USB keydisk, astkd should equal */dev/sda1*. To use another partition on the system, just fill in the path to that partition.

Hardware Requirements

To use AstLinux, you will need at least the following:

- A Soekris net4801/PC Engines WRAP board (net4801 image) or any modern PC hardware with a Pentium or better processor
- A CompactFlash IDE adapter (i586 image)
- A USB CF adapter (or an IDE adapter)
- A computer already running Linux or Windows
- A CompactFlash card of 32 MB or larger (a 256 MB SanDisk is recommended)
- A PC with two Ethernet devices (one is acceptable, as discussed shortly)

Install from Windows

I have tried very hard to make AstLinux as easy to install and configure as possible. The simplest way to get started is to go to the AstLinux web site at *http://www.astlinux.org/* and look for the Downloads section. Once there, you should find the Windows install package, which you will want to download and save to a local disk. Once you run the install package, follow the prompts until it notifies you of a successful installation. Under the Programs group in the Start menu, you should see a new entry called AstLinux, with shortcuts to creating CFs and some documentation.

Attach your USB CF adapter (with CF inserted) and click on the shortcut for the image that you would like to create. A screen will appear, prompting you to select a target disk. This is the harder part of the install, because many people don't know one disk from the other. What I can tell you is that it is usually the last disk listed, but I cannot be sure because all machines are different. One thing to note is that by default, the CF writing utility will refuse to write to disks larger than 800 MB. This will prevent you from accidentally overwriting your hard disk (which should be much larger than 800 MB).

Once you have selected your disk, follow the remaining prompts until you see the progress counter write the entire image and the words "Press any key to continue" appear. You can now safely remove the USB adapter.

Although your fresh, new CF is now ready to be used, I would like you to take a look through the AstLinux User Guide, which is also available from the AstLinux Programs group. Because AstLinux was created from scratch, it bears little resemblance to any existing distributions, and the User Guide attempts to familiarize the user with its features and configuration.

Install from Linux

As stated earlier, go to the Downloads section of *htttp://www.astlinux.org/*. Here you should find compressed (gzipped) versions of the AstLinux images. Download the image you would like and save it to a place on your hard drive. Connect your USB CF adapter (with CF inserted), and look in */var/log/messages* to see what device it was assigned. If you don't have any other USB or SCSI disks attached to the system, it should be located in */dev/sda*. That is what I will assume, but make sure to note whether your Compact-Flash card is located in a different device. To verify the location of your CF card, type the following:

```
# fdisk -l /dev/sda
```

You should see the partition table and drive layout information for your CF card. Now it's time to burn the image. At the command prompt as root, type the following:

```
# gunzip -c /path/to/imagefile.img.gz > /dev/sda
```

where */path/to/imagefile.img.gz* is where you downloaded the image file to and */dev/sda* is where your CF card is located. After the command completes and you are returned to the shell prompt, you can remove your USB

CF writer. As with my Windows installs, I highly recommend that you read the AstLinux User Guide. Because you didn't download a package, you should go back to *http://www.astlinux.org/* and download the User Guide to familiarize yourself with AstLinux.

Install from CD-ROM

The newest way to install AstLinux is via a more traditional means: an install from CD-ROM. As with the other versions of AstLinux, you can download the install CD-ROM image from the Downloads section of *http://www.astlinux.org/*. Once you have downloaded the ISO image, you can write it to a CD-ROM under Windows using such tools as Nero. Under Linux, *cdrecord* (or a graphical frontend such as *K3b*) works quite well.

Once you have written the ISO image, insert the CD in the drive of your soon-to-be-AstLinux machine. Make sure that the machine is set to boot from CD-ROM, and power on. Once the machine boots, you should see a very simple instruction screen. Typing **install** and pressing Enter will start the AstLinux installer. It will attempt to detect any hard drives in the system and prompt you as to which one you would like to install to. You should choose your selection carefully, as AstLinux will overwrite any data on that disk!

> Actually, the install portion will also detect USB CF writers as hard disks (*sda, sdb, sdc*, etc.). This way, you can boot the machine from the CD and write to an AstLinux Compact-Flash card without ever touching the machine's hard drive!

Don't install at all! While someone who is serious about setting up an AstLinux/Asterisk server would not use this method in production, the same AstLinux CD-ROM image used in the preceding section can also be used as a Live CD. This is actually the default! Once you have created the CD, you simply boot the machine and accept the defaults. Hopefully, in a matter of moments, you will be running AstLinux without having to overwrite your hard drive. And hopefully you will like what you see and decide to run the installer as mentioned earlier in this hack!

More about the AstLinux CD-ROM. The AstLinux CD-ROM also includes a nifty Windows Autorun portion that will give you access to the AstLinux User Guide, a link to the web site, and the tools and utilities provided by the Windows install package. Try it out!

Boot Time

After reading the User Guide, it's time to boot! Insert the CF, make sure that the machine will boot from the CF, and power on! After POST, you should see the GRUB menu with a few options available. For now, it's probably best to select the first entry. By default, AstLinux will attempt to obtain an IP address via DHCP on the first Ethernet interface that it finds, and it will statically configure the second interface with an RFC 1918 private address to do Network Address Translation (NAT). If this is not optimal for your situation, I will show you how to change this once the system boots.

After the usual kernel messages go by, you should finally get to a login prompt; log in with the username **root** and the password **astlinux**. Now that you are logged in, it's time to set up your system.

The first thing you are going to want to do is set up your keydisk. As mentioned before, a keydisk is a separate partition or device that AstLinux will use to store your configuration. I am going to assume that you are using a second Flash drive (such as a USB pen drive) for a keydisk and that the USB Flash drive is the Linux device */dev/sda*. (When you boot with the CF disk, it is considered an IDE device and should appear on your system as */dev/hda*— so, don't think I'm asking you to overwrite your CF disk.) Verify that Linux can see the keydisk by typing the following:

```
# fdisk -l /dev/sda
```

You should see the partition table for your device. Make sure to take a good, hard look at it, because now is the time to tell you that in a matter of moments, we will be erasing everything on that device! If this is not OK, remove the USB drive and chose another one. If it is OK to lose all of the data on this Flash drive, move on to the next paragraph.

Now that we have those warnings out of the way, let's finally create your keydisk by typing the following:

```
# genkd
```

The *genkd* script will take care of finding the device, partitioning it, formatting it, and copying some base configuration files to it. You should see some status information and messages go by, but it should be finished in no time, returning you to the command prompt. If you would like to verify it was successful, type the following:

```
# ls /mnt/kd
```

You should see a file there called *rc.conf*. If it's there, you should now type **reboot** to restart the system and begin using the keydisk. If it's not there, make sure that your device really is */dev/sda*, and that it is connected, etc.

Once the system has booted back up, you can start making configuration changes. The *etc/rc.conf* file is where you are going to want to begin to look. In an attempt to keep AstLinux small, I've included only the *vi* text editor, as it was part of the BusyBox collection of utilities I was already using (*http://www.busybox.net/*). If you're not comfortable editing text files with *vi*, you can use the web interface. To use the web interface, make sure that you are using a machine on the internal interface of the AstLinux machine (eth1) and that you have obtained a DHCP address from the AstLinux machine. Once you have done this, simply point your web browser to *https://pbx* (this resolves only if you got your DHCP information from the AstLinux machine, so be sure you are connected to the right interface). You will be prompted for a username (admin) and password (astlinux). Go to General, then Setup, and then "Edit rc.conf." The *rc.conf* file should open up in a small text edit window inside your browser. Make any necessary configuration changes, including setting the EXTIP family of variables. To apply these changes, simply save the file and reboot the system. In future versions of AstLinux, so many reboots won't be necessary, but for now it is always nice to know that the system will come back up after you have made your changes.

PBX-Only Mode (or Help! I Have Only One Ethernet Interface!)

As noted earlier, you don't really need two Ethernet interfaces. If you don't want to use AstLinux as a router, you need only one interface when you configure AstLinux for *PBX-only* mode. PBX-only mode will prevent AstLinux from attempting to configure your internal interface (eth1), and it will prevent the startup of certain services that are not necessary (*iptables*, routing, QoS, DHCP, etc.).

You can configure PBX-only mode by commenting out INTIF= in *rc.conf* and rebooting. Note that the configuration for EXTIF still applies as usual to your first Ethernet interface, eth0.

Wrap-Up

After the system boots, you should verify IP connectivity. You can do this by using the ping command to attempt to reach a remote system. So, try typing **ping www.google.com**. You should see ping replies. If you do not, you might be having Internet issues, or you might have to configure a static IP address.

If the ping is successful, you have correctly set up AstLinux! Feel free to log into the system through the console or SSH and take a look around. Explore the web interface, as a lot of neat things are happening there. If you have any questions, you can always go to the AstLinux-Users mailing list at *http://lists.kriscompanies.com/*. Enjoy!

—*Kristian Kielhofner*

Build a Standalone Voicemail Server in Less Than a Half-Hour

If you're a decent typist, it might take you only 15 minutes.

Asterisk comprises many quality applications, and voicemail is one of them. In fact, Asterisk is perhaps best known for the feature set of its voicemail system. In this section, I will demonstrate how you can harness Asterisk's extremely powerful voicemail application in 30 minutes or less. This way, everybody in your house (or your office) can have a customized voicemail greeting and message recorder, even if they don't have a desktop phone. Think of this as a road-warrior voicemail solution.

First, you will need to download my Asterisk distribution, AstLinux. AstLinux is made to run from CompactFlash, but it doesn't have to be that way. (Incidentally, if you'd like to run it from CompactFlash, check out "Build a SoftPBX with No Hard Drive" [Hack #95]. Pay attention to the keydisk portion, as you'll want one for this hack!) But because we want our voicemail server to be as reliable as possible, I am going to assume that you have a fair amount of storage space—CompactFlash or hard disk—available for use on this hack.

To build a standalone voicemail server, you'll need the following:

- A standard PC with AstLinux
- An IDE-to-CF adapter
- A CompactFlash card of 32 MB or greater (a 256MB SanDisk is recommended)

Depending on what type of technology you are going to integrate this server with, your hardware needs will vary. If you are looking for an all-VoIP solution (possibly for use with another SIP PBX/proxy, etc.), you won't need any additional hardware, and you can skip ahead to the actual setup.

However, if you will be interfacing with a legacy PBX, you will need to get yourself some PSTN interface hardware. For some options, visit *http://www. digium.com/* and *http://www.sangoma.com/*. The PCI interface cards from Digium and Sangoma allow you to connect to a legacy PBX using POTS lines or T1s. (For a crash course in configuring a Digium TDM card, read "Connect a Legacy Phone Line Using Zaptel" [Hack #44].)

Now for the nitty-gritty. After you have AstLinux running (and have made a keydisk), you need to do away with the default Asterisk configuration. There is just too much there for this simple task. You can blow it away by

using this simple command (oh, and if there's anything in this file you want
to keep, back it up first):

```
# echo > /etc/asterisk/extensions.conf
```

You then need to add some basic meat back into *extensions.conf*. Open
extensions.conf in a text editor (*vi* is included by default) and add the
following:

```
[general]
static=yes
writeprotect=yes
autofallback=no

[globals]
VMBASE=8XXX

[default]
include => vmserv

exten => i,1,Hangup
exten => t,1,Hangup

[vmserv]
exten => _${VMBASE},1,Voicemail(u${EXTEN}@vmserv)
exten => _${VMBASE},2,Hangup

exten => _9${VMBASE},1,VoicemailMain(${EXTEN:1}@vmserv)
exten => _9${VMBASE},2,Hangup
```

The first two lines under [general] tell Asterisk never to overwrite this file
with something you tell it dynamically. This is a good idea. The next line
tells Asterisk never to try to guess what to do if no action is assigned. For
this simple configuration, it won't make much difference, but it is generally
a good idea.

The line under [globals] is what you will want to pay the most attention to.
Here, we are setting a variable named VMBASE that will contain the value of
our mailboxes. In this configuration, we are creating a range of extensions
that will map into a range of voice mailboxes. At this point, that range is
8000–8999 (8XXX). If this does not match what you have or need, change it
now, as we will be using this variable throughout this hack.

Underneath [default], we are telling Asterisk to include the separate section [vmserv]. We are also defining what to do when a call goes to an invalid
extension or times out: hang up on them!

The [vmserv] context is where the magic happens. We are using the variable
${VMBASE} to create a range of extensions. We are also telling Asterisk that

when we get a call for one of those extensions, we should put that call into the voice mailbox of that extension, which has the same number. We will play back the unavailable greeting from that mailbox and hang up on the caller when he is finished leaving a message.

So how do we retrieve these voicemails? Simple; all you have to do is call into the Asterisk system and add a 9 before your mailbox number. So if your mailbox number is 8000, extension 8000 will allow callers to leave a message in mailbox 8000. To check mailbox 8000, you will call extension 98000. There are many ways to do this, and I suggest that you look into Asterisk substrings and *extensions.conf* to get a better idea. But for now, save *extensions.conf* because we are done here.

Create the Voice Mailboxes

Now that we have told Asterisk what to do with incoming calls, we need to tell Asterisk what voice mailboxes we want. The voicemail application is configured with the file—you guessed it—*voicemail.conf*. As we did before, open it with *vi* or the web editor.

In the [general] section, uncomment forcename = no and set it to **forcename = yes**. This option enables Asterisk to force a new user to record his real name when he first accesses his voicemail. Asterisk determines whether a user is new by his PIN. If his PIN and voice mailbox are identical, Asterisk will guide him through setting up his voice mailbox. Scroll down to the bottom of *voicemail.conf*, and create a new section that looks like this:

```
[vmserv]
8000 => 8000,Lisa Hayes,lisa@rt.com
8001 => 8001,Rick Hunter,rick@rt.com
8002 => 8002,Lynn Minmei,lynn@rt.com
8003 => 8003,Max Sterling,max@rt.com
8004 => 8004,Miriya Sterling,miriya@rt.com
```

Here, you are creating five voice mailboxes for a fictional group of five folks. The fields in *voicemail.conf* map out like so:

```
mailbox number => PIN,Real Name,E-mail address
```

There are many more options, but you will have to dig deeper into Asterisk on your own time to discover them. You have only 30 minutes to get this done for me to be true to the title of this hack!

Now, all that remains is actually creating the directory structure for the mailboxes. AstLinux includes the *addmailbox* script from the *contrib/scripts* directory of the Asterisk source code. It is extremely easy to use and will do all the work for you based on your input. At a shell prompt, simply type `addmailbox` to get started. It will ask you for the voicemail context. Enter `vmserv`. It will then ask you for the mailbox number. Enter **8000**. Congratulations! You just gave Lisa Hayes a mailbox. To create mailboxes for the rest of the users, simply rerun *addmailbox*, replacing 8000 with 8001, 8002, and so on.

Final Setup

If you would like to use the voicemail-to-email functionality provided by Asterisk and AstLinux, you will need to edit the */etc/rc.conf* file and fill in the variable `SMTP_SERVER` with the IP address or hostname of an SMTP server that will relay mail for your Asterisk server. If you are delivering email to only one domain name, you can use the SMTP server for that domain name, as it will accept mail for it from any system. After making any other configuration changes, you will want to reboot your server to verify its configuration.

Now, how are you going to get your callers into this system? If you have a SIP platform, you simply need to send the callers into the PBX with a simple SIP URL, such as *8000@<IP Address of Asterisk server>*. This SIP URL will put the caller into Lisa Hayes's mailbox.

If you are using PSTN hardware (POTS/T1/E1/etc.), you are going to need to set up Zaptel and Zapata [Hack #44]. The same principles as before still apply; just make sure that you are sending callers into the default context when they ring in on Asterisk's Zaptel channels. And there you have it: a state-of-the-art voicemail system using all open source components, done in 30 minutes or less.

—*Kristian Kielhofner*

H A C K
#97

Automate Your Voicemail Greeting
Use an AGI script with Asterisk to update your voicemail greeting automatically.

In the business world, people often update their voicemail greeting on a daily basis. For example, you might call a co-worker and be greeted with "You've reached the desk of Bob Smith. Today is Tuesday, August 16, and I am in the office today." You can imagine Bob's routine when he gets into the office in the morning: he verifies the current date, rehearses his new message a couple of times, and then calls into his voicemail and updates it. He probably stumbles over the words a couple of times, so he probably has to start over at least once.

This takes far more effort and time than I am willing to commit just to update the current date in my voicemail greeting. I've got more fun things to work on than that! Here's how you can use an AGI script with Asterisk and some home automation to keep your voicemail greeting up to date automatically, without lifting a finger.

Create the Sound Files

You can always use Asterisk's built-in text-to-speech engine to speak your message for you, but that is a little too cold. Instead, with a little work, you can have Asterisk play the appropriate sound files that you've recorded with your own voice. I used the sound recorder that comes with Windows to record several sounds: one for each day of the week (*wday1.wav* for Monday, *wday2.wav* for Tuesday, etc.); one for each month of the year (*month1.wav* for January, *month2.wav* for February, etc.); and one for each day of the month (*1.wav* for the first, *2.wav* for the second, etc.). You'll also want to record a beginning for your greeting, and two different endings, one for when you're in the office and one for when you are out. I named these files *start.wav*, *endnormal.wav*, and *endooo.wav*.

> When speaking these months and days, you might be surprised how difficult it is to get the words to flow together without sounding choppy. I finally started saying an entire date and then cropping the file at the appropriate place to get it to flow better when Asterisk plays it.

By default, Asterisk doesn't have a codec to play *.wav* files. Instead of installing a new codec, you can use the SoX sound converter [Hack #24] (*http://sox.sourceforge.net/*) to convert the files into the *.gsm* format that Asterisk can play with its default installation. Use the following command to convert a *.wav* file into a *.gsm* file:

```
$ sox foo.wav -r 8000 foo.gsm resample -ql
```

After you've converted all your sound files, create a directory called *vm-sounds* in */var/lib/asterisk/sounds* and copy the files into it.

Motion Detection Code

I set up a motion detector right under my desk in my office. I use an excellent home automation package called MisterHouse (*http://www.misterhouse.net*) to monitor the motion detector. I know that if there is no motion in my office between 8 a.m. and 9 a.m., I am probably not going to be in the office that day. Here's the code file I give to MisterHouse to write a file if I'm in the office between those times:

```
$office_movement = new X10_Item('A1'); # A1 is the X10 code

my $office_presence_start = "8:00 AM";
```

```perl
my $office_presence_end = "9:00 AM";
my $office_presence_file = "office.presence.txt";

if (time_now($office_presence_start)) {
   # reset the file
   unlink $office_presence_file;
}

if (state_now $office_movement eq ON
   and time_greater_than($office_presence_start)
   and time_less_than($office_presence_end)) {
   open PRESENCE, ">$office_presence_file";
   print PRESENCE time();
   close PRESENCE;
}
```

Save the previous code in a file and place it in the MisterHouse code directory. You'll need to reload MisterHouse to have it start using your code.

Dialing Greeting Code

Here is the code I use to control the creation of my dialing greeting. Note that it requires the Asterisk::AGI Perl module:

```perl
#!/usr/bin/perl
use strict;
use Asterisk::AGI;

my ($sec,$min,$hour,$mday,$mon,
   $year,$wday,$yday,$isdst) = localtime(time);
my $yyyymmdd = sprintf ("%04d%02d%02d",$year+1900,$mon+1,$mday);
my $month = $mon + 1;

my $vm_sound_dir = "vm-sounds";
my $office_presence_file = "/path/to/file/office.presence.txt";

my $AGI = Asterisk::AGI->new;
my %input = $AGI->ReadParse;

my $greeting_start = "start";
my $greeting_end_in_office = "endnormal";
my $greeting_end_out_of_office = "endooo";
my $weekday_sound = "wday" . $wday;
my $mday_sound = $mday;
my $month_sound = "month" . $month;

my @files_to_play = ();
push @files_to_play, $greeting_start;
push @files_to_play, $weekday_sound;
push @files_to_play, $month_sound;
push @files_to_play, $mday;
```

```
if (-e $office_presence_file or $hour == 8) {
        push @files_to_play, $greeting_end_in_office;
} else {
        push @files_to_play, $greeting_end_out_of_office;
}

foreach my $sound (@files_to_play) {
        $AGI->verbose("Playing sound $sound");
        $AGI->stream_file("$vm_sound_dir/$sound");
}
```

Save the code to a file named *vmautomate.pl* and place it in the */var/lib/asterisk/agi-bin* directory. Add the following lines to your *extensions.conf* file, where *8001* is your extension and *100* is your voice mailbox number:

```
exten => 8001,1,Dial(SIP/8001,20,rt)
exten => 8001,2,AGI(vmautomate.pl)
exten => 8001,3,Voicemail,100
```

After you reload Asterisk, when you call your extension, your phone will ring for 20 seconds and then the AGI script will run. Depending on the time of the day and the presence of *$office_presence_file*, you should hear the appropriate greeting. You should replace your "regular" voicemail greeting with the default greeting so that it flows properly once it reaches the Voicemail directive.

It doesn't take long to imagine some cool functionality with AGI scripts. For example, you can take this a step further and have an AGI script that reads your calendar that you've published with iCal to see whether you're scheduled to be out of the office that day. You can even have Asterisk announce to a caller when you have free time during that particular day and suggest that the caller try back then.

—Dave Mabe

HACK #98   Connect Asterisk to the Skype Network

They said it couldn't be done without the Skype API. They were wrong.

Wouldn't it be great if your Asterisk server could place calls to (and receive calls from) the Skype network? Imagine the possibilities: putting your Skype buddy list within reach of the Asterisk dial plan so that all your calls can be routed to the appropriate Skype buddy depending on what you dial on your Asterisk-connected phone. Well, that dream is now a reality—with a few gotchas.

The first gotcha is that you'll need to use a little bit of legacy technology (FXO/FXS interfacing) to set up the connection. The second gotcha is that you'll need a Windows PC sitting next to your trusty Asterisk server, and that Windows PC will need to have Skype running and an Internet Phone Wizard USB interface attached. The final gotcha is that the Internet Phone Wizard must already have speed-dial numbers associated with the members of your Skype buddy list [Hack #40]. You would need this anyway if you were going to use the Internet Phone Wizard for its intended purpose: connecting a traditional analog phone to the Skype network via a Skype client on the USB host PC.

Connect a standard RJ11 phone cord from the telephone jack on the Internet Phone Wizard to an FXO port on your Asterisk box. This FXO port can be on a Digium TDM400P, a Digium X100P, or a Sangoma WANPIPE PCI card (several examples for setting up the TDM400P are given in Chapter 4). Configure the Zaptel channel for this FXO port as you normally would if you wanted to connect the Asterisk server to a standard phone line. (The standard phone line is going to be substituted by the connection from the Internet Phone Wizard.)

Next, you'll need to add your Internet Phone Wizard speed-dial numbers to your Asterisk dial plan so that they'll be dialed via the Zaptel FXO channel. (Remember, I'm assuming you've already set up your speed-dial numbers on Skype using the Internet Phone Wizard, so if you haven't, flip back to "Skype with Your Home Phone" [Hack #40].) In the default Asterisk context for the phone you're going to be calling from, add something like this:

```
Exten => 71,1,Dial(Zap/1/71)
Exten => 72,2,Dial(Zap/1/72)
Exten => 73,3,Dial(Zap/1/73)
```

If you had buddies with speed-dial numbers of 71, 72, and 73, the Asterisk server would attempt to call them on Skype via the connected Internet Phone Wizard. Of course, all the Asterisk box sees is the Zaptel interface. To be even slicker, you can assign an entire range of numbers to be used for Skype purposes. Here, I've set aside 80 through 89. The dial plan will always dial these extensions on the Zaptel interface, passing the extension number through to the Internet Phone Wizard as dialed digits:

```
exten => 8X,1,Dial(Zap/1/${EXTEN})
```

To get calls from Skype routed into your Asterisk dial plan, all you need to do is modify the default context of the Zaptel channel you've used to connect the Internet Phone Wizard. Refer to "Connect a Legacy Phone Line Using Zaptel" [Hack #44] for an example that points out how to do this. Now, if you really want to get fancy with Asterisk and Skype, check out the next hack.

Forward Your Home Phone Calls to Skype

HACK #99

For those times when you really, really need to stay in touch.

This is just plain cool. If you've come this far with Asterisk and the Internet Phone Wizard, you've unlocked a world of wicked-cool hack potential. To get you primed for your journey to Aster-Skype hackatopia, let me show a very simple dial-plan modification that will simultaneously ring your incoming phone calls on your locally connected phones as well as on your Skype phone. In */etc/asterisk/extensions.conf*, consider the following:

```
exten => s,1,Dial(SIP/100&SIP/200)
```

This extension will dial the two phones connected on SIP peers 100 and 200, and connect the call to whichever peer answers first. But let's say an Internet Phone Wizard is connected to channel Zap/2 [Hack #98]. Now, you can actually dial those two phones and a Skype speed-dial alias from your buddy list:

```
exten => s,1,Dial(SIP/100&SIP/200&Zap/2/99)
```

Now, whoever is associated with speed-dial number 99 in your Skype buddy list will also receive a call (through the Skype network), courtesy of the Internet Phone Wizard that you've hooked up to a Zaptel FXO port (Zap/2) on the Asterisk machine. So, if this was your default context for incoming calls from your home phone line (connected to Asterisk via another channel), your incoming home phone calls would also ring on the Skype client that was logged in as the buddy in your list.

Ideally, this buddy is a second Skype account you've set up, because Skype doesn't let you "call yourself." So, I would set up Ted1 and Ted2 as Skype buddy names and then have the Internet Phone Wizard's Skype client log in as Ted2, while I go about my normal business logging in as Ted1. In essence, my home phone calls will be forwarded from Skype user Ted2 to Skype user Ted1.

Get Started with sipX

HACK #100

Asterisk, like Cisco CallManager and other softPBX platforms, implements SIP as a method of supporting SIP phones and trunks, but does not employ the SIP design philosophy. Yet SIP and SIP alone can replace your entire PBX system. Enter sipX.

Like Asterisk, the sipX project implements a call-management server for Linux, implements a voicemail server with message-waiting indicators, and allows you to build a voice network of SIP phones. Unlike Asterisk, sipX does it exclusively using SIP. This means that external interface gateways must be used to communicate between sipX and non-SIP networks (the PSTN, H.323, etc.). In a minute, at least if you follow this little outline, you'll be installing sipXpbx, a comprehensive SIP PBX server.

SipXpbx brings some cool functionality to the table, including a built-in web-based administration tool, two SIP softphones (sipXPhone and sipXez-Phone), and a suite of interoperability testing tools. Awesome stuff! Perhaps most important, sipX implements the following components of a SIP network according to the official IETF SIP specifications (unlike Asterisk, which only implements certain parts of a SIP network):

Registrar

A SIP server that keeps track of SIP clients by tracking the IP addresses where they're located and the usernames associated with each registration; a directory of active SIP clients, if you will

Proxy

A server that relays SIP messages and media streams between disparate networks

Client

A user agent that uses SIP for telephony, text messaging, voice chat, or some other media applications

sipX's Requirements

sipX runs on Linux. There's presently some limited support for BSD, but Windows and Mac users, at least for the moment, are out of luck. An installer is available for Fedora Core 3, a distribution of Linux put forth by Red Hat. In fact, Fedora Core 3 is an ideal environment for sipX; 256 MB of RAM and 500 MB of available disk space are plenty for setting up a sipX test lab.

When setting up Linux for sipX, be sure to install the PostgreSQL database as well as the Apache Web Server, both of which sipX utilizes.

Install sipXpbx

Get logged on to your target machine as root. I'm going to assume that you're running Fedora Core 3. For other distributions, such as Fedora Core 2 and Gentoo Linux, see the official sipX compatibility list at *http://www.sipfoundry.org/*. First, download and run the sipX Fedora Core 3 install script:

```
# wget http://www.sipfoundry.org/pub/sipX/sipXpbx-2.8.1-fc3.sh .
# sh ./sipXpbx-2.8.1-fc3.sh
```

When prompted by the script, answer **y** and be sure to enter a password for the `sipxchange` user that the script creates. That's all there is to installing sipXpbx.

Next, you've got to generate an SSL certificate for sipX to use:

```
# mkdir $HOME/sslkeys
# cd $HOME/sslkeys
# /usr/bin/ssl-cert/gen-ssl-keys.sh
# /usr/bin/ssl-cert/install-cert.sh server-01
# /usr/bin/ssl-cert/install-ssl-keystore.sh server-01
```

Use the default password of `changeit`, and answer **yes** when the script asks whether you trust the certificate.

Launch sipXpbx

Starting and stopping sipXpbx is a snap. Use the *service* command like this to start sipX, and replace `start` with `stop` to stop it:

```
# service sipxpbx start
```

If sipXpbx complains about HTTPD syntax errors the first time you try to launch it, just give your Fedora machine a reboot.

Finish sipXpbx Setup by Web Interface

sipX uses the JBOSS application server as the foundation of its excellent web-based GUI. To access the web-based site installation wizard in your web browser, visit the following URL, replacing *sipx.your.domain* with the address of your sipX server:

```
http://sipx.your.domain:8080/pds/ui/install/install.jsp
```

On the web page that appears, enter **installer** as the username and **password** as the password. Then click continue, and you'll be greeted by the sipX Configuration Server, as shown in Figure 7-7.

Here, you'll want to enter your friendly organization name, your server's DNS domain name, a simple authentication realm name (of your own choosing), and an alphanumeric PIN that will later serve as your administrator password on sipX. Click Submit, and after a few moments, you should be looking at the standard login prompt.

Use the username **superadmin** and the password you established on the PIN prompt in the preceding page. Click Submit, and watch as sipXconfig loads the administrative GUI.

This is where extra RAM and a fast processor will really come in handy. Unlike Asterisk and the Asterisk Management Portal, which use Perl and

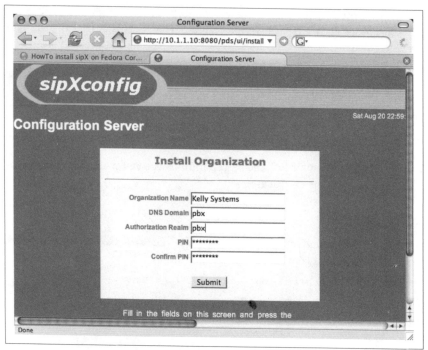

Figure 7-7. sipX's initial configuration screen

PHP and run swiftly on a minimally configured machine, sipXconfig is a highly sophisticated set of Java applications that really call for heavy-duty server hardware. A Pentium 4 PC with at least 512 MB of RAM should be sufficient to run sipXconfig at an acceptable pace.

Register for the Administration Guide

To expand your sipX prowess beyond installation, you should get your eyes on the sipX Administration Guide. But to do this, you've got register with the chief commercial sponsor of the sipX project, Pingtel. The URL for registration is *https://secure.pingtel.com/registration/registerUser.jsp*.

Index

Numbers

3Com's 48-volt IntelliJack switch converter, 170
3D Avatar Messenger, 90
802.1p, 15
  support by standalone Ethernet switches, 16
802.1p precedence tagging, 189
  checking support with pathping, 190
802.3af standard for inline power, 169
8-bit pulse code modulation formats, 57
8x8 Inc.'s DTA-310, standard equipment for Packet8 service, 4
911 emergency service, 7, 17–19, 128
  compromise solutions, 18
    POTS line, keeping, 18
    public safety dispatcher number, 19
    speed dialing with VoIP device, 18
    using a cell phone, 19
  problems with VoIP emergency dialing, 17
  TSPs and, 17

A

access signaling by analog devices, 9
access signaling protocol (SIP), 9
ACK method (SIP), 224, 225
Actiontec Internet Phone Wizard, 92

Address Book, calling contacts with Skype, 70–72
ADSL subscribers using Linux, 203
aLaw encoding, 57
Ambrosia Software
  Snapz Pro X, 61
  WireTap Pro, 55
AMP (Asterisk Management Portal), 134–137
  configuring MySQL database, 136
  downloading and installing, 135
  Perl modules, installing, 135
  running install script, 137
  setup process, 134
  software prerequisites, 134
analog modems, devices using to communicate on phone lines, 9
analog phones and phone lines, connecting to Mac mini PBX, 263
analog telephone adapters (see ATAs)
analog telephones, 2
  connecting to Asterisk server, 95
  connecting to SPA-2000 ATA, 4
  hooking up to Skype network, 92–94
  placing and receiving calls via the Internet, 3
animated, three-dimensional emoticons, 90
Answering Machine, Skype, 84–87
  rotating greetings with Windows, 86
AOL Instant Messenger (AIM), 26

We'd like to hear your suggestions for improving our indexes. Send email to *index@oreilly.com*.

Apache, 284
    use by AMP web-based GUI, 134
Apple
    AirPort Express, 79
    Dashboard widget system, 24
    iLife applications, 40
    Mac mini, 262–265
Apple iChat, support for SIP, 28
AppleScript
    caller ID for Phlink, 44
    calling Address Book contacts with
        SKype, 70–72
application-based QoS, 15
ARP (Address Resolution Protocol), 238
ARP poisoning, 238–240
Asterisk, 95–149, 241
    attaching SIP phone, 100–104
        configuring BudgeTone
            101, 101–102
        enabling IP phone to place
            calls, 102–104
        setting IP phone to use SIP
            server, 102
        testing Asterisk, 104
    connecting legacy phone line using
        Zaptel, 108–112
    connecting PSTN phone line using
        Sipura SPA-3000
        ATA, 210–212
    connecting standard phone line using
        FXO gateway, 105–108
    connecting telephony devices
        to, 95–97
        FXO or FXS interfaces, 96
        T1, using, 96
    connecting to Skype network, 281
    faxes, handling, 254–256
    forwarding home calls to your cell
        phone, 112–114
    four-line phone server,
        building, 124–128
        setting up incoming calls, 127
        setting up outgoing calls, 128
        setting up station-to-station
            calls, 127
    getting daily weather forecast, 133
    installing and testing server on Linux
        PC, 97–100
        Linux-specific start and stop
            scripts, 100
        starting and stopping server, 99

integrating X10 controls with phone
    system, 179–181
LDAP client, 257
linking serveral PBXs over the
    Internet, 142–145
    adding remote locations, 144
    configuring dial plan, 144
linking servers with PSTN, 138–142
    configuration, 140–141
    controlling caller ID, 141
Mac mini PBX, 263–265
monitoring from Perl
    scripts, 266–268
music-on-hold, 129–131
    assigning different phones/lines to
        different music classes, 131
    streaming MP3 Internet radio
        station, using, 131
PrivacyManager, using, 122–123
recording calls, 131–132
reporting telephone activity with
    Excel, 116–121
routing calls with distinctive
    ring, 145–147
running without root, for
    security, 137
sample set of configuration files, 99
selectively forwarding calls, 114–116
software components for Linux, 98
teaching Asterisk box to
    speak, 259–262
tuning up logs, 147–149
    changing default storage
        location, 148
    size of logfiles, 148
    syslog as target for output, 149
voicemail greeting updates,
    automating, 278–281
voicemail server, building, 275–278
web-based administration
    interface, 133–137
Asterisk Gateway Interface, 128
Asterisk Management Portal (see AMP)
Asterisk Manager API, 128
    commands, 268
asterisk (modular software daemon), 98
Asterisk::AGI Perl module, 280
Asterisk::Manager class, 268
asterisk-perl module, 266
    Asterisk Manager example, 267

AstLinux, 269–274, 275
  addmailbox script, 277
  booting, 273
  current features, 269
  hardware requirements, 270
  installing from CD-ROM, 272
  installing from Linux, 271
  installing from Windows, 270
  keydisk, 270
  PBX-only mode, 274
AstShape Provider script, 207–210
AstShape script, 202–206
ATAs (analog telephone adapters), 2
  automatic registration with VoIP
    service provider, 5
  broadband routing and firewall
    functions, 4
  choosing your own, 9
  connecting analog phones to Mac
    mini PBX, 263
  connecting to your network, 3
  connection to modular phone
    jack, 7, 8
  keeping firmware up to date, 19–21
  listing of ATAs that provide, 3
  media gateway, 105
  number of phones connected, 9
  placement of, 4
  prioritizing traffic from, 16
  providing dial tone to your analog
    phones, 7
  rebooting, 19
  Sipura ATAs
    building a bat phone, 162–165
    tweaking, 157–162
  Sipura SPA-3000, 210–212
audio blogging (see podcasting)
audio chat, recording on a Mac, 55–57
audio fidelity, phone calls on a
  softphone, 55
Audio Hijack, 55
Audio Voice Cloak, 29
Authentication ID, AIP endpoint on
  Asterisk server, 107
authentication, SIP, 10, 14
Authorization User setting (X-Lite), 13
auto-attendant for calls (Call Soft
  Pro), 39
automatic call answering, Asterisk
  server, 108

automatic ring-through, 163–165
Avantlook, 70
avatars (buddy icons), Gizmo, 32
Away, Busy, or Available status
  indicators, 73

B

base stations (extended range) for
  wireless LANs, 79
bat phone (automatic ring-through),
  building, 162–165
blogging, audio blogging (see
  podcasting)
Bring Your Own Device (BYOD) service
  agreement, 9
broadband router, configured as DHCP
  server, 2, 5
broadband routing in ATAs, 4
broadband VoIP service, 9
BroadVoice
  support of configurable
    softphones, 11
  web-based tool to place and
    manipulate calls, 23
Broadvox Direct VoIP service
  web-based toolset to configure
    find-me-follow-me call list, 24
buddy list (Skype), putting Jyve into, 74
BudgeTone 101 IP phone, 101–104
  configuring, 101
Busy, Away, or Available status
  indicators, 73
BYE method (SIP), 226

C

cable modems, voice traffic over, 201
cables
  connecting ATA to your network, 3
  RJ11-equipped telephone patch
    cable, 7
Cacophony (sound-editing tool), 52
  using in podcasting, 82
Cain & Abel, 234–237
  ARP poisioning used to intercept
    call, 238–240
call detail records (CDRs), reporting
  with Excel, 116–121
call events, standards for, 27

call forwarding
    from phone company lines to
        Asterisk server, 107
    home phone calls to Skype, 283
    home phone calls to your cell
        phone, 112–114
    selectively forwarding calls on
        Asterisk, 114–116
call quality, maximizing, 14–17
call sniffer/recorder, Cain & Abel, 234
Call Soft Pro application, 39
Call411 application, 38
caller ID, 115
    controlling for Asterisk PBX when
        using PSTN trunks, 141
    having Asterisk server to
        announce, 262
    identifying calls without, 122
    in Phlink AppleScripts, 44
    pop-up caller ID notification
        (Phlink), 41
    prompting user to enter, 123
    Windows software for, 36
${CALLERIDNUM} channel
    variable, 114
call-out credits (Gizmo), 34
calls
    audio fidelity of calls on
        softphone, 55
    handling with Windows
        software, 36–39
    recording VoIP calls on Windows
        PC, 35
    recording VoIP calls with Gizmo, 33
    sampling resolution for typical phone
        calls, 52
CDRs (call detail records), reporting
    with Excel, 116–121
cell phones
    forwarding home calls to, 112–114
    using for 911 service, 19
CG-200 and CG-400 media
    gateways, 105
channel bank, 97, 108
channels, 103
    Aseterisk voice channel,
        defining, 109
    call-forwarding setup using two SIP
        peers, 114
    FXO, corresponding to VoIP
        channels, 107

    FXO, setting up with Asterisk, 109
    legacy interface ports on TDM400P
        card, 126
chorus effect (SoX), 58
Cisco IP phones
    customizing boot logo, 171
    powering with standard inline
        power, 169
    speed-dial service, 166–169
Cisco switches, PortSPAN
    technique, 237
classes, music-on-hold, 131
Click-2-Call (Vonage), 23
clients
    Gizmo, downloading and
        installing, 31
    Trillian instant-messaging client, 34
clients, VoIP, 26–29
    communication protocols, 27
    comparison of features and
        compatibility, 28
    H.323 clients, 27
    IAX (Inter-Asterisk Exchange), 28
    SIP (Session Initiation Protocol), 27
    understanding features, 28
Clipcomm CG-200 and CG-400 media
    gateways, 105
CM15A controller, 180
codecs
    compatibility for SIP
        phones, 222–226
    G.711, 232
    PreferredCodec SIP setting, 176
    processor power and, 244
    selection, BudgeTone IP phone, 152
    supported by Uniden IP phone, 175
command line interface (CLI),
    Asterisk, 104
command-line interface (CLI),
    Asterisk, 99
CompactFlash-based PC, using for
    softPBX, 268–274
compatibility, 243
compression, voice chat tools, 84
conference calls
    Skype, recording for podcast panel
        discussion, 83
    Skype users, triggering from the
        Web, 77
    TSP policies on, 10

congestion
  signaled by Asterisk, 144
  source of jitter and latency, 194
Contact header, 197
contact search, Skype, 66
contexts (Asterisk), 111
  dialing out on legacy phone lines
      connected to Asterisk
      PBX, 128
  editing for TDM400P-connected
      phone lines, 127
  incoming, 113
CounterPath
  web site, user manual for X-Lite, 14
  X-Lite (see X-Lite softphone)
country codes for international calls, 80
credentials, media gateway, 106
cron utility, 43
curl utility, 25, 133
  required for AMP, 134
currency conversions, 80
CVS repository for Asterisk (at
      Digium), 97

D

Dashboard widget system (Apple), 24
Dataprobe AutoPAL, 20
date format for Asterisk logs, 148
demarc (telephone company entry
      point), 7
  disconnecting wires from phone
      company, 8
desktop telephony, 22–62
  Audio Voice Cloak, 29
  choosing VoIP client, 26–29
  creating telephony sounds with
      SoX, 57–59
  getting info with Google, 50
  Gizmo Project, 31–35
  handling calls with Windows
      software, 36–39
  Mac, answering and logging your
      calls, 40–48
  mixing sound files for announcement
      message, 59
  professional-sounding greetings, 60
  recording audio chat on Mac, 55–57
  sound files, telephony-ready, 52

tracking Vonage account info, 24–26
VoIP, using with online
      games, 48–50
DHCP
  IP address assignment to ATAs, 163
  problems with changes in dynamic IP
      addresses, 19
  turning off for BudgeTone IP
      phone, 152
DHCP server, broadband router
      configured as, 2, 5
Dial command, 123
dial plan
  Asterisk, 112
      commands, 115
      for PBXs linked over the
          Internet, 144
      Monitor command, 132
      routing Skype calls to, 282
  Sipura ATA, 160–162
      automatic ring-through, 164
dialing by IP address, 153
dialing shortcuts for pure VoIP, 5–7
  listed, 6
DIDs, resolving to user email
      address, 257–259
DiffServ (QoS mechanism), 175, 228
digital telecommunications (T1), 96
digital telephones, recording calls on
      your PC, 151
Digium, 95
  CVS repository for Asterisk
      software, 97
  IAXy FXS gateway, 181
  TDM400P card, 124–127
      installing, 125
  Wildcard TDM400, 182
  X100P FXO card, 165
Direct Inward Dial, 139
DirectX, 48
Display Name setting (X-Lite), 12
distinctive ring, 39, 145–147
dmesg command, 231
DNS Address setting for TSP provider
      (X-Lite), 12
DNS records for Asterisk servers, 143
DNS server
  Grandstream BudgeTone IP
      phone, 152
  Uniden IP phone, 173

Index | 291

Do Not Call Registry, 50
Domain/Realm setting (X-Lite), 13
downsampling, 52
   recordings of audio chats, 55
   with SoX utility, 59
dropped packets, 14
DSL
   avoiding accidental disconnection, 8
   voice traffic over, 201
DTMF Dial widget, 183
dynamic IP addresses, changes in, 19

E

echo
   on Ventrilo voice chat system, 48
   with softphones, 14
echo command output, flowing to
      Festival standard input, 262
echo test service, Skype, 67
effects, adding with SoX, 58
eight-wire CAT5 patch cable with two
      RJ45 connectors, 3
electrical damage to the ATA from
      phone-company lines, 8
email
   routing faxes into, 256–259
   voicemail-to-email functionality, 278
email integration, AMP, 135
emergency 911 calls, 128
emergency 911 service, 7, 17–19
   compromise solutions, 18
      POTS line, keeping, 18
      public safety dispatcher
         number, 19
      speed-dialing with VoIP
         device, 18
      using a cell phone, 19
   problems with VoIP emergency
      dialing, 17
   TSPs and, 17
emoticons, using with Skype, 89–92
   adding sound and video
      emoticons, 90–91
   real-time video, 91
encodings, sound files, 57
   mixed voice and music file for
      announcement, 60
equalization
   SoX effect, 58
   voice chat tools, 84

Ethereal, 215–222
   capturing failed capabilities
      negotiation, 224–226
   capturing SIP statistics, 220
   capturing successful capabilities
      negotiation, 223
   configuring, 216
   observing SIP registration, 217
   SIP registration failure,
      observing, 220
   sniffing out jittery calls, 226–228
Ethernet connections, IP phones
      powered over, 169
Ethernet interface, ATAs, 3
Ethernet port, 2
Ethernet switches
   prioritizing traffic from, 16
   ToS feature, 15
euros, converting to other currencies, 80
extended-range base station for wireless
      LAN, 79
extensions.conf file, 112
   context for SIP phones dialing out on
      Asterisk PBX with legacy
      phone lines, 128
   context section, editing for
      TDM400P-connected phone
      lines, 127
   dial plan for SIP peer incoming
      context, 113
   music-on-hold, 130
   PrivacyManager command, 123

F

fast-busy, 144
fax calls, answering with Phlink, 42
fax machine
   turning Linux box into, 254–256
   using analog modems over traditional
      phone lines, 9
faxes, routing automatically into emails
      and PDF files, 256–259
fax/modem card, converting to Zaptel
      interface card, 165
fax-receiving support for AMP, 136
Festival (speech synthesizer), 260–262
   reading weather report, 133
Festoon, 91

FIFO (first in, first out) queues,
      manipulation by QoS
      technology, 202
file format conversion (sound files), Sox
      utility, using, 57
find-me-follow-me call list, 24
Firefly, 28
      support for SIP and IAX, 27
firewall functions in ATAs, 4
firewalls
      IAX and, 28
      NAT, 197
      NetFilter, logging VoIP
            traffic, 228–232
      optimizing local firewall on
            softPBX, 246
      SIP and, 63
Flash storage device, IP phones, 166
foreign exchange office (see FXO)
foreign exchange station (see FXS)
forums, Skype, 66
four-wire patch cable with two RJ11
      connectors, 4
FWD (Free World Dialup), 6
FXO (foreign exchange office), 96
      channel, setting up with
            Asterisk, 109
      Digium X100P FXO card, 165
      gateway connecting phone line to
            Asterisk, 105–108
      gateway device connecting analog
            phones to Mac mini PBX, 263
      interface card, configuring to use FXS
            signaling, 110
      interface card, installing, 109
      modifying FXO channel for pulse
            dialing, 182
      modules on TDM400P card, 126
FXS (foreign exchange station)
      Digium IAXy FXS gateway, 181
      interfaces, 96
      Kewlstart signaling, 126
            configuring for FXO interface
                  card, 110
      modules on TDM300P card, 126
      pulse dialing and, 183

G

G.711 codec, 232
games, VoIP, using with, 48–50

GarageBand, using in podcasting, 82
gateways, 96
      Digium IAXy FXS gateway, 181
      (see also media gateways)
      (see also router address)
Gauge It (Gizmo Project), 33
genkd script, 273
Gizmo Project, 22, 31–35
      conferencing limitations, 49
      downloading and installing client, 31
      extra features, 33
            call-out credits, 34
            Gauge It, 33
            Map It, 33
            Record It, 33
      placing a call, 32
      support for SIP, 28
Global System for Mobile, 48
GnoPhone, 27, 241
      support for SIP, 28
GNU Gatekeeper (gnugk), 249
GnuGK, 241
Google
      getting telephony info, 50
      images, 172
Google Groups, 52
Google Talk, 50
GotoIf command, 115
Gotta Go widget, 67
Grandstream BudgeTone
      making IP-to-IP phone
            calls, 151–155
      mounting on the wall, 153
Grandstream IP phones, custom
      ringtone, building, 155–157
graphical user interfaces (GUIs)
      Asterisk, 134
      web-based sipX GUI, 285
graphing jitter and latency data
      (RRDtool), 194–196
greeting messages
      different message each day, 43
      mixing different sound files for, 59
      professional-sounding, 60
      rotating for SAM with Windows, 86
      voicemail, updating
            automatically, 278–281
greeting script, 41, 43
GSM (Global System for Mobile)
      codec, 48, 57, 224

H

H.323 standard, 27
  building H.323 gatekeeper using
    OpenH323, 247–253
  VoIP clients, 27
Hangup command, 123
hangup script, 47
hardening a server, 245
  cleaning up xinetd, 245
  optimizing local firewall on
    softPBX, 246
  removing unnecessary software, 245
hardphones (IP), 151
hardware, telephony, 150–183
  bat phone (automatic
    ring-through), 162–165
  configuring multiple IP phones at one
    time, 172–177
  controlling house lights from IP
    phone, 179–181
  creating a Zaptel interface card, 165
  custom ringtone for Grandstream
    phone, 155–157
  customizing Cisco IP phone boot
    logo, 171
  customizing Uniden IP phones from
    TFTP, 177–179
  IP-to-IP phone calls with
    Grandstream
    BudgeTone, 151–155
  powering Cisco phones from
    non-Cisco switches, 169
  recording calls from standard phone
    on a PC, 150
  Sipura ATA, tweaking, 157–162
  speed-dial service on Cisco IP
    phones, 166–169
  using rotary-dial phone with
    VoIP, 181–183
headers, SIP message packets, 188
headers, SIP packets, 10
headphones, for voice communication
  with gaming, 50
high availability (telephony server), 243
high priority for voice media traffic, 15
historical newsgroup search tool
  (Google Groups), 52
HotRecorder 2.0 for Windows, 90
HP iPAQ hx4700 Pocket PC, 78
HylaFAX, 256

I

IAX (Inter-Asterisk Exchange)
    protocol, 27, 144
  register feature with dynamic IP
    addresses, 143
  support by JAJAH, 34
  used with legacy signaling on
    Asterisk PBX, 112
  VoIP clients, 28
IAXy FXS gateway, 181
iChat, 28
ICMP packets, sending with
    traceroute, 191
ID3 library, 129
iLife applications, 40
images, available on Google images, 172
IMTO: prefix, 76
incompatible phones, failed calls
    from, 222
inline power, adding to CAT5/CAT6
    cable connection, 170
inline power (standard), for Cisco
    phones, 169
inline recorder switches, 150
instant messaging, 26
  Gizmo Project, peer-to-peer
    softphone, 31
  JAJAH, 34
  recording voice calls on Windows
    PC, 35
  Skype instant messaging, adding to
    web site, 76
Intel 537EP chipset, 165
Intel V.92 Data/Fax/Voice modem
    card, 165
interactive voice response (IVR), 57
  Sipura ATA, 157
Inter-Asterisk Exchange protocol (see
    IAX protocol)
interface cards, FXO, 109
international calls, SkypeOut, 79
Internet
  routing calls over using pure VoIP
    dialing, 5–7
  using to link several remote Asterisk
    PBXs, 142–145
Internet Phone Wizard, 92, 282, 283
Internet telephony service providers (see
    TSPs)
INVITE method (SIP), 188, 222

IP addresses
    assignment to ATAs, 163
    BudgeTone IP phone, 152
    dynamic, changes in, 19
    locating your ATA IP address, 21
    setting for Uniden IP phones, 172
IP (Internet Protocol), User Datagram
        Protocol (see UDP)
IP phones, 3, 9
    allowing to place calls via
        Asterisk, 102
    Cisco
        customizing boot logo, 171
        powering from non-Cisco
            switches, 169
        speed-dial service, 166–169
    configuration with TFTP, 213–215
    configuring BudgeTone 101, 101
    configuring multiple IP phones at one
        time, 172–177
    controlling house lights
        from, 179–181
    custom ringtone for Grandstream
        phone, 155–157
    customizing Uniden IP phones from
        TFTP, 177–179
    Grandstream BudgeTone 101,
        mounting on the wall, 153
    making IP-to-IP phone calls with
        Grandstream
        BudgeTone, 151–155
    recording calls on your PC, 151
    setting to use SIP server, 102
IP Precedence, 228
IP telephony, 1
IP telephony access devices, 9
iptables, 231
    policy commands for kernel
        firewall, 247
IP-to-IP calling, 153
    enabling for Uniden IP phone, 176
iTunes
    controlling from Phlink, 46–48
        pausing and resuming for
            calls, 47
        web site for scripts, 48
IVR (interactive voice response), 57

J

Jabber protocol, 50
JAJAH, 34
JavaScript, pop-under window sending
        Skype instant message on visit
        from Jyve user, 76
JBOSS application server, 285
jitter, 193
    buffer settings for Uniden IP
        phone, 175
    determining from graph of latency
        data, 196
    measured by traceroute, 192
    network congestion as source of, 194
    sniffing out with Ethereal, 226–228
jitter buffers, 193
Journal feature (Outlook), 69
Jyve, 73–75
    creating account, 74
    HTML to embed Q-Card in your web
        page, 74
    making a Skype buddy, 74
    text messages, sending, 74
    web browser plug-in, 76–77

K

Kewlstart signaling (FXS), 110, 126
keydisk (AstLinux), 270, 273
Konfabulator (see Yahoo! Widgets)

L

language setting, Uniden IP phone, 176
LARTC, 206
latency, 193, 203
    graphing over time with
        RRDtool, 194–196
    measuring on a route, 191
    network congestion as source of, 194
LDAP inquiry to associate DIDs with
        email address, 257–259
LDAPget package, 257
libdnet library, compiling and
        installing, 233
libevent library, compiling and
        installing, 233
libpri module, 98
libtiff, 134

line number (Asterisk voice
        channel), 110
Linksys BEFSR81 broadband router,
        prioritizing packets on, 15
Linux
    ADSL subscribers using, 203
    finding TFTP server, 214
    NetFilter firewall, logging VoIP
        traffic, 228–232
    PBX that communicates with the
        PSTN, 108–112
    Skype, 65
    turning into an Asterisk
        PBX, 97–100
        Linux-specific start and stop
            scripts, 100
        starting and stopping Asterisk
            server, 99
    turning Linux box into fax
        machine, 254–256
    voice server OS, selecting, 244
    Zaptel driver framework, 124
local public safety dispatcher, phone
        number for, 19
logging
    configuring for Asterisk, 147–149
        changing default storage, 148
        size of logfiles, 148
        syslog as target for output, 149
    VoIP traffic, 228–232

M

MAC addresses, 238
    Pocket PCs, 79
    Uniden IP phone, 173
Mac OS X
    Cacaphony sound-editing tool, 52
    calling Address Book contacts with
        Skype, 70–72
    installing vonageGauge, 25
    podcasting tools, 82
    recording videoconference with
        Snapz Pro X, 61
    Skype, 49, 65
    Skype Events options dialog, 89
    Soundflower, 31
    Unit Converter widget, 80
    widgets, 24
        installing Yahoo! Widgets, 24

Macintosh
    answering and logging your
        calls, 40–48
    Asterisk server caller ID
        announcement using Say
        command, 262
    building PBX with Mac
        mini, 262–265
    recording audio chat, 55–57
    SoX utility, 57
MAD (MPEG Audio Decoder), 129
madplay command, 130
Map It (Gizmo Project), 33
masquerading, 197
Master.csv file, 116
media gateways, 96
    FXO gateway connecting standard
        phone line to
        Asterisk, 105–108
methods (SIP), 188
mewencode command, 256
microphones
    for Skypecasts, 84
    voice communication with
        gaming, 50
Microsoft Excel, utilization records for
        Asterisk PBX server, 116–121
MIME Construct package, 135
MisterHouse, 279
mixing sound files with SoX, 59
modem card, converting to Zaptel
        interface card, 165
modular phone jacks, 7
    ATA connection to, 8
Monitor dial-plan command, 132
MP3s
    Asterisk music on hold, 129
    editing for telephony with
        Cacophony, 53
    recording on Windows PC, 35
    streaming Internet radio station for
        music on hold, 131
MPEG Audio Decoder (MAD), 129
Mpg123 player, 129
music
    downsampling sound files for
        telephony, 53
    mixing with spoken greeting
        message, 59

music-on-hold, 46
    Asterisk, 129–131
        assigning different phones/lines to
            different music classes, 131
        streaming MP3 Internet radio
            station, using, 131
    downsampling sound files for
        telephony, 53
musiconhold.conf file, 130
MySQL, 134
    CDR interface for Asterisk, 136
    configuring database for Asterisk
        CDRs and AMP, 136

N

NAT (Network Address Translation)
    exploring NAT traversal, 196–201
    STUN protocol, use by IP
        phones, 176
National Do Not Call Registry, 50
National Oceanic and Atmospheric
    Administration (NOAA), 133
native Voice over IP, 6
ncurses, 134
NetFilter, logging VoIP traffic, 228–232
NetMeeting, 27
    registering H.323 endpoint, 251–253
Net::Telnet Perl module, 135
NOAA (National Oceanic and
    Atmospheric
    Administration), 133

O

OhPhone, 241, 247
OhPhoneX, registering H.323
    softphone, 250
on-hold music (see music-on-hold)
online forums, Skype, 66
online gaming, VoIP, using with, 48–50
OpenGK, 247
OpenH323, 241, 247–253
    downloading and compiling
        OpenH323 Gatekeeper
        (gnugk), 249
    installing, 248
    software packages, 247
OpenLDAP, 257
OpenMCU, 248
OpenSSL, 134

operating system, selecting and
    hardening for voice
    server, 244–247
OPTIONS method (SIP), 187, 188
Outbound Proxy setting (X-Lite), 13
Outlook contacts, Vonage calls to, 23
Outlook, placing Skype calls
    from, 67–70
Ovolab, Phlink scripting, web site, 48

P

packet interval, 175
packet jitter, 14
Packet8, firmware patches for the
    ATA, 21
packets
    prioritizing
        received packets and, 16
    prioritizing to improve VoIP
        quality, 14–17
panel discussion, using recorded Skype
    conference call, 83
Party Line (Gizmo Project), 32
Password setting (X-Lite), 13
pathping tool, 189–191
PBXs
    Asterisk server supporting four legacy
        phones, 124–128
    building Mac PBX, 262–265
    linking several Asterisk PBXs over the
        Internet, 142–145
        adding remote locations, 144
        configuring dial plan, 144
PC expansion (PCI) cards, 95
PC softphones, 9
PCI interface cards, 275
PCMA codec, 152
PCMU codec, 152
PDF file, converting faxes to, 259
peers
    configuring IAXy for use as, 182
    SIP, configuring for media
        gateway, 108
peer-to-peer (P2P) network, Skype, 63
Perl, 134
    Grandstream-compatible ringtone
        script, 155–157
    installing modules for Asterisk, 135

Perl (*continued*)
  script investigating how SIP
    works, 187
  script monitoring SIP hosts on VoIP
    system, 184–187
  script recording latency and jitter
    data, 192–194
  scripts to monitor Asterisk
    PBX, 266–268
Phlink, 40–48
  answering fax calls, 42
  caller IDs in AppleScripts, 44
  controlling iTunes, 46–48
    pausing and resuming for
      calls, 47
  custom greetings, 41
  greeting callers differently each
    day, 43
  pop-up caller ID notifications, 41
  running when you are logged off, 42
phone calls (see calls)
phone jacks, modular, 7
phone numbers, finding with
    Google, 50
  searching for your own phone
    number, 51
phone service providers, 1
phone service, traditional, 9
phone-company lines, disconnecting at
    demarc, 8
PhoneTray Dialup application, 37
PhoneTray Free application, 37
PHP
  for AMP GUI, 134
  modifying configuration files for
    AMP music-on-hold, 135
ping utility, 144
  latency measure, flaw in, 194
  measuring latency, 193
pitch effect, SoX, 61
Pivot Table Report (Excel), 119–121
plain old telephone service (POTS) line
    for 911 calls, 18
Plug and Play (PnP), 10
Pocket PC version of Skype, 65
Pocket PC, WiFi-enabled, using as
    Skype portable phone, 78
PocketSkype, 78

podcasting
  integrating Skype into, 81–84
    experimenting for perfect
      Skypecast, 83
    Mac podcasting tools, 82
    recorded conference calls, 83
    Windows podcasting tools, 82
  Mac OS X tools for, 82
Polycom IP500 phone, 179
port (default) for SIP, 153
port numbers
  used by VoIP applications,
    controlling, 207
  VoIP, commonly used, 230
Port SPAN, 237
port spanning, 216
ports
  SipPort setting, Uniden IP
    phone, 174
  used for VoIP protocols, 246
PostgreSQL database, 284
post-processing tools for podcasts, 84
POTS (Plain Old Telephone Service) line
  connecting to FXO interface
    card, 109
power injector, 170
power over Ethernet (PoE), 169
power supplies, redundant, 243
PowerDsine, 170
presence, 73
primary rate interface (PRI), 96
prioritizing packets, 14–17
  on Linksys broadband router, 15
  received packets and, 16
  RTP traffic, 15
privacy check for your phone
    number, 51
PrivacyManager command, 123
proprietary signaling protocol,
    Skype, 63
Provider DNS Address setting
    (X-Lite), 12
proxy servers, SIP settings, 13, 174
PSAP (Public Safety Answering
    Point), 128
PSTN (Public Switched Telephone
    Network)
  avoiding with pure VoIP dialing, 6
  interface hardware, 275

PSTN (*continued*)
interfacing your phone line with VoIP
network, 210–212
linking Asterisk servers
with, 138–142
controlling caller ID, 141
dial-plan configuration, 140–141
Linux PBX that communicates
with, 108–112
placing/receiving calls to and from
Skype network, 63
routing calls from trunk in Asterisk
using distinctive ring, 146
PSTNgw, 248
Public Safety Answering Point
(PSAP), 128
public safety dispatcher, phone number
for, 19
pulse code modulation formats,
8-bit, 57
pulse dialing, 181
pass through, 183
support by Digium IAXy FXS
gateway, 181
support by Digium Wildcard
TDM400, 182
using DTMF Dial widget on
Windows PC or Mac, 183
pure or native Voice over IP, 6
pure VoIP, 95
pure VoIP dialing between TSPs, 5–7
PWLib libraries, 248

Q

Q-Cards (Jyve), 73
creating and customizing, 74
embedding in your web page, 74
sending text messages, 74
QoS (Quality of Service), 15
auditing for VoIP network, 189–192
graphing latency and
jitter, 192–196
NAT traversal, 196–201
using pathping, 189–191
using traceroute, 191
defined, 202
DiffServ, 175
implementing where jitter
originates, 228

improving by shaping network
traffic, 201–206
prioritizing some types of network
traffic, 194
WAN interface for, 208
quality of Internet connection to phone
network (Gizmo), 33
quality of VoIP calls,
maximizing, 14–17

R

radio broadcasting, voice recording
techniques, 83
RAID 5 disk array, 243
Real-time Transport Protocol (see RTP)
rebooting the ATA, 19
Record It (Gizmo Project), 33
recorder applications, download
sites, 35
recorder switches, 150
recording
audio from VoIP conversations on
Mac OS X, 82
a videoconference, 61
recording calls
Asterisk, 131–132
audio chat on your Mac, 55–57
from standard telephone on a
PC, 150
with Gizmo, 33
on your Windows PC, 35
Recording Industry Association of
America (RIAA), 81
recording software, generating
telephony-ready sounds, 52
redundant power supplies, 243
Register setting (X-Lite), 14
registrar, SIP, 174
registration
H.323 endpoint, using
NetMeeting, 251–253
H.323 softphone, using
OhPhoneX, 250
SIP, turned off for IP-to-IP
dialing, 153
registration, SIP client, 106
registration, SIP, observing with
Ethereal, 217
Rejection Hotline, 23

remote phone jukebox, 46
requests, SIP, 188
residential-style, single-line phone,
    connecting ATA to, 4
Resource Reservation Protocol (see
    RSVP)
responses to SIP methods, 188
restart commands (Asterisk), 100
reverb effect (SoX), 58
reversal effect (SoX), 58
RFC 3581, 197
ring script, 45
    pausing iTunes for calls, 47
ringtones, custom
    for Grandstream phone, 155–157
ringtones, customized (for Skype), 87
Rogue Amoeba's Audio Hijack, 55
root user account, running Asterisk
    without, 137
rotary-dial phones
    for bat phone, 164
    using with VoIP, 181–183
router (default gateway) address
    BudgeTone IP phone, 152
    Uniden IP phone, 173
    (see also gateways)
RRDtool, 194–196
RSVP (Resource Reservation
    Protocol), 189, 228
    checking support with pathping, 191
RTP (Real-time Transport
    Protocol), 194
    control of port numbers, 207
    NAT and, 199
    prioritizing RTP traffic, 15

S

s extension, 114
S518 ADSL board from Sangoma
    Technologies, 203
SAM (Skype Answering
    Machine), 84–87
    rotating greetings with Windows, 86
sample set of Asterisk configuration
    files, 99
sampling resolution for phone calls, 52
SaRP, 241
Say command, 262
Scheduled Tasks (Windows), 86

SDP (Session Description
    Protocol), 222–226
    inspecting failed capabilities
        negotiation, 224–226
    inspecting successful capabilities
        negotiation, 223
search engine, Google.com, 50
security
    Asterisk, running without root, 137
    global VoIP trunks, 142
    operating system for voice
        server, 244
    Skype network, 65
    wireless LAN, 79
sendmail, 116
services (Cisco), 167
Session Initiation Protocol (see SIP)
shift (pitch) effect, SoX, 61
show application command, 257
signaling
    FXO/FXS devices, 110
    legacy telephony devices on Asterisk
        PBX, 126
    pulse signaling, 181
signaling protocols, 2
    Skype, 63
Simple Mail Transfer Protocol
    (SMTP), 116
Simple Traversal of UDP NATs (see
    STUN)
SIP Express Router, 241
SIP peers
    configuring for media gateway, 108
    connected to the Asterisk server, 113
SIP phone, attaching to
        Asterisk, 100–104
    configuring BudgeTone
        101, 101–102
    enabling IP phone to place calls via
        Asterisk, 102–104
    setting IP phone to use SIP
        server, 102
    testing Asterisk, 104
SIP phones, station-to-station calls, 127
SIP Proxy setting (X-Lite), 13
SIP (Session Initiation Protocol), 9, 27
    calling a party without using SIP
        gateway or gatekeeper, 181
    configuration of Uniden IP
        phone, 173–177

SIP (*continued*)
  controlling port numbers, 207
  default port, 153
  examining SIP packets with
      Ethereal, 215–222
  firewalls and, 63
  inspecting SIP message
      structure, 187–189
  monitoring SIP hosts over VoIP
      network, 185–187
  packet sent through router with SIP
      translation enabled, 198
  poor NAT traversal capabilities, 197
  setting IP phone to use SIP
      server, 102
  use by VoIP clients, 26
  VoIP clients, 27
  X-Lite configuration settings, 11–14
SIP Uniform Resource Indicator
    (URI), 181
sip.conf file, two SIP peers connected to
    Asterisk server, 113
SIPphone Inc., web site, 31
sip_ping.pl script
  exploring NAT traversal, 198
  recording latency and jitter
      data, 192–196
  server responding to host sending
      packet instead of host in
      Contact header, 197
Sipura ATAs
  building a bat phone, 162–165
  tweaking, 157–162
    configuration options via
        IVR, 157
    dial plan, 160–162
    top 10 options, 159
    web interface, 158
Sipura SPA-2000 ATA, 4
Sipura SPA-3000 ATA, connecting PSTN
    phone line to VoIP
    network, 210–212
sipX, 241, 283–286
  finishing sipXpbx setup by web
      interface, 285
  implementation of components of a
      SIP network, 284
  installing sipXpbx, 284
  launching sipXpbx, 285

  registering for Administration
      Guide, 286
  requirements, 284
sipXphone, 28
Skype, 22, 26, 63–94
  calling Mac OS X Address Book
      contacts, 70–72
  capabilities and limitations of, 64
  connecting Asterisk server to, 281
  contact search function, 65
  custom rings and sounds,
      using, 87–89
  download site, 65
  emoticons, using, 89–92
    adding sound and video
        emoticons, 90–91
    real-time video, 91
  forwarding home phone calls to, 283
  hooking up analog phone to, 92–94
  how it works, 64
  ignoring or ending calls, 67
  integraging into podcasts, 81–84
    experimenting for perfect
        Skypecast, 83
    Mac podcasting tools, 82
  integrating into podcasts
    recorded conference calls, 83
    Windows podcasting tools, 82
  Jyve social networking
      service, 73–75
  Jyve web browser plug-in, 76–77
  logging calls with Outlook Journal
      feature, 69
  placing calls from Outlook, 67–70
  placing/receiving calls to and from
      regular phones via PSTN, 63
  Pocket PC version, 65
  putting Skype Me link on your web
      site or blog, 72
  SAM (Skype Answering
      Machine), 84–87
    rotating greetings with
        Windows, 86
  searching user directory for
      contacts, 66
  security, 65
  signaling protocol, 63
  SkypeOut international calls, 79
  testing operation of, 67

Skype (*continued*)
    using WiFi-enabled Pocket PC as
        portable phone, 78
    using with gaming, 49
    Voicemail service, 87
Skype Me mode, 66
Snapz Pro X, 61
sniffing or capturing packets with
        Ethereal, 215–222
Snom 200 SIP hardphone, 188
softPBX, building with no hard
        drive, 268–274
softphones, 2, 3, 27
    audio fidelity of calls, 55
    echo, dealing with, 14
    Gizmo Project, 31–35
    JAJAH, 34
    recording conversations with
        WireTap, 55
    using with a VoIP TSP, 9–14
        differing policies, 10
        installing X-Lite, 10–14
sound codecs, recording and converting
        with Total Recorder, 35
sound emotes, 90
sound files
    converting between formats with
        SoX, 57
    converting with SoX, 88
    creating for voicemail greetings, 279
    mixing different kinds for
        announcement, 59
    telephony-ready, generating, 52
    uLaw, for Grandstream
        ringtone, 155
Sound Recorder, 52
sound recorders, 60
sound-editing tools, 52
Soundflower, 31, 82
SourceForge, VoIP projects, 241
SoX (SOund eXchange), 57–61, 82
    adding sound effects, 58
    converting stereo WAV to mono, 88
    converting .wav file to .gsm, 279
    deepening your voice on greeting
        messages, 61
    mixing recorded Asterisk calls, 132
    mixing sound files for announcement
        message, 59
    required for AMP, 134

resampling and re-leveling
    sounds, 59
sound-file format conversion, 57
spandsp package, 136, 254
    sending faxes, 255
speech guidelines for
    professional-sounding
    announcements, 60
speech synthesis, using with
    Asterisk, 259–262
speed-dial service on Cisco IP
    phones, 166–169
Speex codec, 49
SSL support, curl utility, 25
stability (telephony server), 242
standalone Ethernet switches,
    prioritizing traffic on, 16
standards for signaling call events, 27
standards, supported by Gizmo VoIP
    network, 35
station-to-station calls, SIP phones, 127
stepper switch, 181
stop commands (Asterisk), "now" and
    "when convenient"
    arguments, 100
streaming media devices, interaction
    of, 9
streaming MP3 radio station for music
    on hold, 131
Strowger, Almon B., 181
STUN (Simple Traversal of UDP
    NATs), 176, 197
    building STUN server, 199
subnet mask
    BudgeTone IP phone, 152
    Uniden IP phone, 173
supernodes (Skype), 64
SVMTO: prefix, 77
switches, Cisco PoE, substituting on
    Cisco IP phones, 169
syslog database file, 231
System( ) command (Asterisk), 261

T

T1, 96
tail -f command, 146
tcpdump, 232–234
TCP/IP listeners, eliminating to harden a
    voice server, 246

TCP/UDP port numbers, packet priority
based on, 15
TDM (time division multiplexing), 124
64Kbps PCM bitstream codecs that
mimic, 175
dedicated time slots for calls, 201
Digium Wildcard TDM400, 182
interface cards, 142
TDM400P card, 124–127
installing, 125
team voice chat system (Ventrilo), 48
Teamspeak (for gaming), 49
telemarketer calls, handling on
Asterisk, 122–123
telephone company's point of entry (see
demarc)
telephones
analog phones, 2
IP phones, 3
softphones, 3
wiring your house phones for
VoIP, 7–9
telephony hardware, 150–183
bat phone (automatic
ring-through), 162–165
configuring multiple IP phones at one
time, 172–177
controlling house lights from IP
phone, 179–181
creating a Zaptel interface card, 165
custom ringtone for Grandstream
phone, 155–157
customizing Cisco IP phone boot
logo, 171
customizing Uniden IP phones from
TFTP, 177–179
IP-to-IP phone calls with
Grandstream
BudgeTone, 151–155
powering Cisco phones from
non-Cisco switches, 169
recording calls from standard
telephone on a PC, 150
rotary-dial phone, using with
VoIP, 181–183
Sipura ATA, tweaking, 157–162
speed-dial service on Cisco IP
phones, 166–169
telephony info, finding with Google, 50

telephony server, building, 242–247
H.323 gatekeeper, using
OpenH323, 247–253
procesor power needed, 243
selecting and hardening an
OS, 244–247
three areas of focus, 242
telephony service providers (see TSPs)
telephony-ready sound files,
generating, 52
Telnet interface, IP phones, 102, 152
text expressions for happiness and
dissatisfaction, 89
text message to answer Skype calls, 86
text messages, Jyve, 74
TFTP Desktop, 213
TFTP servers
configuration files for IP phones, 171
customizing Uniden IP
phones, 177–179
Grandstream IP phone, 157
IP phone configuration, 213–215
mass-configuring Uniden IP
phones, 172–177
setting address for, 175
SIPDefaults.cnf file, editing to set
Services URL, 167
TIFF file, fax image saved to, 255
tiff2ps utility, 255, 259
time division multiplexing (TDM), 124
time shifting effect (SoX), 58
Time::HiRes Perl module, 185
TiVo boxes, ATAs and, 9
ToS (Type of Service), 15, 190
support by standalone Ethernet
switches, 16
using to prioritize class of voice
traffic, 207, 209
TOSC, Call Soft Pro, 39
Total Recorder, 35
adjusting sound resolution and
output format, 35
Total Recorder Standard Edition, 82
TR16A phone controller, 179
traceroute, 189, 191
latency measure, flaw in, 194
traditional phone service, 9
traffic-shaping script
(AstShape), 202–206
transducer pickup, 151

Trillian (instant-messaging client), 34
TSPs (telephony service providers), 2
  desktop-based extras provided
      by, 22
  emergency 911 service, 7, 17
  listed, with web sites, 3
  Provider DNS Address, 12
  SIP proxies, 13
  SIP-based, Skype vs., 64
  updating ATA firmware, 21
  using pure VoIP dialing with, 5–7
  using softphone with, 9–14
two-wire phone splitters, 7
Type of Service (ToS), 15, 190

U

UDP (User Datagram Protocol), 2
  prioritizing packets to improve VoIP
      quality, 14–17
uLaw or aLaw encoding, 57
uLaw sound file, 155
Uniden IP phones
  customizing from TFTP, 177–179
  mass configuring by TFTP, 172–177
Unit Converter widget, 80
Universal Currency Converter, 80
Unix, finding TFTP server, 214
Use Outbound Proxy setting
      (X-Lite), 13
User Datagram protocol (see UDP)
User ID, registering on Asterisk
      server, 106
User Name setting (X-Lite), 12
UTP Ethernet patch cable, hacking to
      plug Cisco IP phones into
      802.3af source, 170

V

V.92 PCI modem chip family, 165
variables, channel, 114
V-Cards, 73
Ventrilo, 48
verbose logging output (Asterisk), 146
Via header, 197
vibrato effect (SoX), 58
video conference, recording, 61
video4IM, 91
video-on-Skype plug-ins, 91
visual emoticons, 90

VLANs (Virtual LANs), 175, 228
voice channels (Asterisk), 109
voice mailboxes, creating, 277
voice recording techniques, 83
voice server, building, 242–247
  H.323 gatekeeper, using
      OpenH323, 247–253
  processor power needed, 243
  selecting and hardening an
      OS, 244–247
  three areas of focus, 242
voice-alteration tools
  Audio Voice Cloak, 29
  Soundflower, 31
voicemail
  Asterisk service, 104
  automating greeting
      updates, 278–281
voicemail alternative for Skype, 77
voicemail and auto-attendant (Call Soft
      Pro), 39
voicemail server (standalone),
      building, 275–278
Voicemail service (Skype), 87
VoicePulse, 4
  service to dump unwanted girlfriends
      or boyfriends, 23
VoIP networks
  auditing QoS capabilities, 189–192
    graphing latency and
        jitter, 192–196
    NAT traversal, 196–201
    using pathping, 189–191
    using traceroute, 191
  building PSTN gateway, 210–212
  creating premium class of
      service, 206–210
  inspecting SIP message
      structure, 187–189
  intercepting and recording a VoIP
      call, 237–240
  IP phone configuration, 213–215
  logging and recording VoIP
      streams, 234–237
  logging VoIP traffic, 228–232
  monitoring VoIP devices, 184–187
  peeking inside SIP packets, 215–222
  SDP (Session Description
      Protocol), 222–226
  secretly recording calls, 232–234

VoIP networks (*continued*)
    shaping network traffic to improve
        QoS, 201–206
    sniffing out jittery calls with
        Ethereal, 226–228
VoIP (Voice over IP)
    choosing desktop client, 26–29
    defined, xiii
    emergency dialing, problems
        with, 17
    getting connected, 2–5
    phone service providers, 1
    programming device to mimic
        911, 18
    telephony service providers, listed, 3
    wiring your house phones for, 7–9
vomit (Voice over Misconfigured
        Internet Telephones), 232,
        234
Vonage
    911 service, 17
    calls to Outlook contacts with one
        click, 23
    tracking account info on your
        desktop, 24–26
    X-PRO commerical counterpart of
        X-Lite, 11
vonageGauge widget, 24
    installing, 25
    using, 26

W

WAV files
    recorded Asterisk calls, 132
    recording on Windows PC, 35
    using as custom Skype ringtones, 87
wcfxo.c file, 165
wcfxs driver, 126
wctdm driver, 125
wctdm module, 109
weather forecast, getting from your
        telephone, 133
web browsers, Jyve plug-in, 76–77
web sites
    simplifying communication for
        visitors, 77
    Skype instant messaging, adding
        to, 76
    tracking visits from Jyve users, 76

web-based call-management tools
        (BroadVoice), 23
web-based GUI, sipX, 285
webcams for Skype real-time video, 91
Weird Solutions, TFTP Desktop, 213
WEP (Wireless Encryption
        Protocol), 79
widgets, 24–26
    DTMF Dial widget, 183
    Gotta Go, 67
    installing Yahoo! Widgets, 24
    Unit Converter, 80
    vonageGauge, 24
        installing, 25
        using, 26
WiFi-enabled Pocket PC, using as Skype
        portable phone, 78
Wildcard TDM400, 182
Windows Mobile 2003, 78
Windows systems
    Audio Voice Cloak, 29
    call-handling software, 36–39
    HotRecorder 2.0, 90
    Jyve web browser plug-in, 76
    Messenger IM software, 27
    pathping tool, 189–191
    podcasting tools, 82
    recording VoIP calls, 35
    Skype, 49, 65
    Skype Answering Machine
        (SAM), 85
    Skype Sound Alerts options
        dialog, 88
    Sound Recorder, 52
    SoX utility, 57
    Trillian instant-messaging client, 34
    widgets, 24
        installing vonageGauge, 25
        installing Yahoo! Widgets, 24
windump, 232
Wireless Encryption Protocol
        (WEP), 79
wireless LAN, security, 79
WireTap Pro, 55
workgroup Ethernet switches,
        prioritizing traffic on, 16

X

X10 phone controller, using to control lights in your home, 179–181
X100P FXO card, 165
X-Lite softphone, 10, 27, 200
  installing, 10–14
    SIP configuration settings, 11–14
  making the call, 14
  recording conversations with WireTap, 55
  SIP packet monitoring with Ethereal, 215–222
  user manual, download in PDF format, 14
XML files, Cisco IP phone menus, 166–169
X-PRO (Vonage version of X-Lite), 11

Y

Yahoo! Chat, Skype vs., 64
Yahoo! Messenger, 26
  support for SIP, 28
Yahoo! Widgets, 24, 81
  installing, 24
  tone dialing using Windows PC or Mac, 183

Z

zapata.conf file, 111
  distinctive ring, 146
  settings of channels provided by TDM400P card, 126
Zaptel driver framework (Asterisk), 96, 105, 124
  connecting legacy phone line to Asterisk, 108–112
  music-on-hold bridging, 130
Zaptel drivers, launching at system startup, 125
Zaptel interface cards, creating your own, 165
Zaptel interface channels, 110
  configuring to detect distinctive signals, 146
zaptel module, 98, 109
Zaptel-compliant interface cards, 98
zaptel.conf file, TDM400P card settings, 126
Zoom 5567 (ATA), 4
ztcfg application, 126
ztdummy driver, 130

Colophon

Our look is the result of reader comments, our own experimentation, and feedback from distribution channels. Distinctive covers complement our distinctive approach to technical topics, breathing personality and life into potentially dry subjects.

The tool on the cover of *VoIP Hacks* is a Morse code tapper. Also known as a telegraph key, this electrical switching device is used to send Morse code over electrical wires.

The old-school variety of telegraph key, glamorized in many classic films, was the straight key, a simple contraption fashioned from a bar with a knob fastened atop one end. When the knob was depressed, the bar completed an electrical circuit, and current flowed through the telegraph wires. By rapidly forming and breaking this circuit, telegraphers could transmit a series of signals, conventionally known as "dits" and "dahs" (or, more colloquially, "dots" and "dashes"), which spurred an electromagnet on the receiving end to produce clicking noises that could be recorded to paper tape or deciphered directly by skilled operators.

Unfortunately, design constraints of the straight key limited its transmission capabilities to a mere 20 words per minute. Additionally, the vigorous "brass pounding" required of early telegraphers sometimes led to a repetitive stress injury called *glass arm*, known today as carpal tunnel syndrome.

Sanders Kleinfeld was the production editor, and Audrey Doyle was the copyeditor for *VoIP Hacks*. Sanders Kleinfeld proofread the book. Philip Dangler and Claire Cloutier provided quality control. Ellen Troutman Zaig wrote the index.

Marcia Friedman designed the cover of this book, based on a series design by Edie Freedman. The cover image is from the Classic Business Equipment CD in the Classic Photographic Image Object Library. Linda Palo produced the cover layout with Adobe InDesign CS using Adobe's Helvetica Neue and ITC Garamond fonts.

David Futato designed the interior layout. This book was converted by Keith Fahlgren from Microsoft Word to Adobe FrameMaker 5.5.6 using open source XML technologies. The text font is Linotype Birka; the heading font is Adobe Helvetica Neue Condensed; and the code font is LucasFont's TheSans Mono Condensed. The illustrations that appear in the book were produced by Robert Romano, Jessamyn Read, and Lesley Borash using Macromedia FreeHand MX and Adobe Photoshop CS. This colophon was written by Sanders Kleinfeld.

Better than e-books

Buy *VoIP Hacks* and access the digital
edition FREE on Safari for 45 days.

Go to www.oreilly.com/go/safarienabled
and type in coupon code GAJS-XPWK-LKCX-UAKQ-PDKK

Search
thousands of
top tech books

Download
whole chapters

Cut and Paste
code examples

Find
answers fast

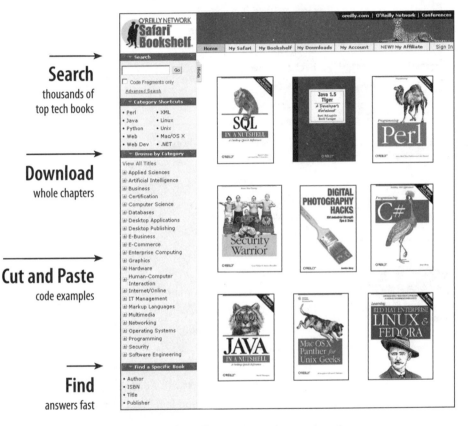

Search Safari! The premier electronic reference
library for programmers and IT professionals.

Related Titles from O'Reilly

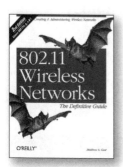

Networking

802.11 Wireless Networks: The Definitive Guide, *2nd Edition*

Asterisk: The Future of Telephony

Cisco Cookbook

Cisco IOS Access Lists

Cisco IOS in a Nutshell, *2nd Edition*

DNS & BIND Cookbook

DNS & BIND, 4th Edition

Essential SNMP, *2nd Edition*

Exchange Server Cookbook

IP Routing

IPv6 Essentials

IPv6 Network Administration

LDAP System Administration

Managing NFS and NIS, *2nd Edition*

Network Troubleshooting Tools

RADIUS

sendmail, *3rd Edition*

sendmail Cookbook

SpamAssassin

Switching to VoIP

TCP/IP Network Administration, *3rd Edition*

Unix Backup and Recovery

Using Samba, *2nd Edition*

Using SANs and NAS

VoIP Hacks

Windows Server 2003 Network Administration

Zero Configuration Networking: The Definitive Guide